LONDON MATHEMATICAL SOCIETY STUDENT TEXTS

1 Introduction to combinators and λ-calculus, J.R. HINDLEY & J.P. SELDIN
2 Building models by games, WILFRID HODGES
3 Local fields, J.W.S. CASSELS
4 An introduction to twistor theory, S.A. HUGGETT & K.P. TOD
5 Introduction to general relativity, L.P. HUGHSTON & K.P. TOD
6 Lectures on stochastic analysis: diffusion theory, DANIEL W. STROOCK
7 The theory of evolution and dynamical systems, J. HOFBAUER & K. SIGMUND
8 Summing and nuclear norms in Banach space theory, G.J.O. JAMESON
9 Automorphisms of surfaces after Nielson and Thurston, A. CASSON & S. BLEILER
10 Nonstandard analysis and its applications, N. CUTLAND (ed)
11 Spacetime and singularities, G. NABER
12 Undergraduate algebraic geometry, MILES REID
13 An introduction to Hankel operators, J.R. PARTINGTON
14 Combinatorial group theory: a topological approach, DANIEL E. COHEN
15 Presentations of groups, D.L. JOHNSON
16 An introduction to noncommutative Noetherian rings, K.R. GOODEARL & R.B. WARFIELD, JR.
17 Aspects of quantum field theory in curved spacetime, S.A. FULLING
18 Braids and coverings: selected topics, VAGN LUNDSGAARD HANSEN
19 Steps in commutative algebra, R.Y. SHARP
21 Representations of Finite Groups of Lie Type, FRANÇOIS DIGNE & JEAN MICHEL

London Mathematical Society Student Texts 21

Representations of Finite Groups of Lie Type

FRANÇOIS DIGNE
University of Picardy at Amiens

and

JEAN MICHEL
DMI, École Normale Supérieure

The right of the University of Cambridge to print and sell all manner of books was granted by Henry VIII in 1534. The University has printed and published continuously since 1584.

CAMBRIDGE UNIVERSITY PRESS
Cambridge
New York Port Chester Melbourne Sydney

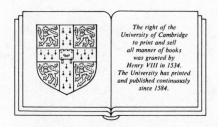

Published by the Press Syndicate of the University of Cambridge
The Pitt Building, Trumpington Street, Cambridge CB2 1RP
40 West 20th Street, New York, NY 10011–4211, USA
10 Stamford Road, Oakleigh, Melbourne 3166, Australia

First published 1991

Printed in Great Britain at the University Press, Cambridge

Library of Congress cataloguing in publication data available

British Library cataloguing in publication data available

ISBN 0 521 40117 8 hardback
ISBN 0 521 40648 X paperback

TABLE OF CONTENTS

INTRODUCTION

These notes follow a course given at the University of Paris VII during the spring semester of academic year 1987–88. Their purpose is to expound basic results in the representation theory of finite groups of Lie type (a precise definition of this concept will be given in chapter 3).

Let us start with some notations. We denote by $\mathbb{F}_q$ a finite field of characteristic p with q elements (q is a power of p). The typical groups we will look at are the linear, unitary, symplectic, orthogonal, ... groups over $\mathbb{F}_q$. We will consider these groups as the subgroups of points with coefficients in $\mathbb{F}_q$ of the corresponding groups over the algebraic closure $\overline{\mathbb{F}}_q$ (which are algebraic reductive groups). More precisely, the group over $\mathbb{F}_q$ is the set of fixed points of the group over $\overline{\mathbb{F}}_q$ under an endomorphism F called the *Frobenius endomorphism*; this will be explained in chapter 3. In the following paragraphs of this introduction we will try to describe, by some examples, a sample of the methods used to study the complex representations of these groups. More examples are developed in detail in chapter 15.

Induction from subgroups
Let us start with the example where $G = \mathbf{GL}_n(\mathbb{F}_q)$ is the general linear group over $\mathbb{F}_q$. Let T be the subgroup of diagonal matrices; it is a subgroup of the group B of upper triangular matrices, and there is a semi-direct product decomposition $B = U \rtimes T$, where U is the subgroup of the upper triangular matrices which have all their diagonal coefficients equal to 1. The representation theory of T is easy since it is a commutative group (actually isomorphic to a product of n copies of the multiplicative group $\mathbb{F}_q^\times$). Composition with the natural homomorphism from B to T (quotient by U) lifts representations of T to representations of B. Inducing these representations from B to the whole of the linear groups gives representations of G (whose irreducible components are called "principal series representations"). More generally we can replace T with a group L of block-diagonal matrices, B with the group of corresponding upper block-triangular matrices P, and we have a semi-direct product decomposition (called a Levi decomposition) $P = V \rtimes L$, where V is the subgroup of P whose diagonal blocks are identity

matrices; we may as before induce from P to G representations of L lifted to P. The point of this method is that L is isomorphic to a direct product of linear groups of smaller degrees than n. We thus have an inductive process to get representations of G if we know how to decompose induced representations from P to G. This approach has been developed in the works of Harish-Chandra, Howlett and Lehrer, and is introduced in chapters 4 to 7.

Cohomological methods

Let us now consider the example of $G = U_n$, the unitary group over $\mathbb{F}_q$. It can be defined as the subgroup of matrices $A \in \mathbf{GL}_n(\mathbb{F}_{q^2})$ such that ${}^t A^{[q]} = A^{-1}$, where $A^{[q]}$ denotes the matrix whose coefficients are those of A raised to the q-th power. It is thus the subgroup of $\mathbf{GL}_n(\overline{\mathbb{F}}_q)$ consisting of the fixed points of the endomorphism $F : A \mapsto ({}^t A^{[q]})^{-1}$.

A subgroup L of block-diagonal matrices in U_n is again a product of unitary groups of smaller degree. But this time we cannot construct a bigger group P having L as a quotient. More precisely, the group $\mathbf{V}$ of upper block-triangular matrices with coefficients in $\overline{\mathbb{F}}_q$ and whose diagonal blocks are the identity matrix has no fixed points other than the identity under F.

To get a suitable theory, Harish-Chandra's construction must be generalized; instead of inducing from $V \rtimes L$ to G, we construct a variety attached to $\mathbf{V}$ on which both L and G act with commuting actions, and the cohomology of that variety with ℓ-adic coefficients gives a (virtual) bi-module which defines a "generalized induction" from L to G. This approach, due to Deligne and Lusztig, will be developed in chapters 10 to 13.

Gelfand-Graev representations

Using the above methods, a lot of information can be obtained about the characters of the groups $\mathbf{G}(\mathbb{F}_q)$, when $\mathbf{G}$ has a connected centre. The situation is not so clear when the centre of $\mathbf{G}$ is not connected. In this case one can use the Gelfand-Graev representations, which are obtained by inducing a linear character "in general position" of a maximal unipotent subgroup (in $\mathbf{GL}_n$ the subgroup of upper triangular matrices with ones on the diagonal is such a subgroup). These representations are closely tied to the theory of regular unipotent elements. They are multiplicity-free and contain rather large cross-sections of the set of irreducible characters, so give useful additional information in the non-connected centre case (in the connected centre case, they are combinations of Deligne-Lusztig characters).

For instance, in $\mathbf{SL}_2(\mathbb{F}_q)$ they are obtained by inducing a non-trivial linear character of the group of matrices of the form $\begin{pmatrix} 1 & u \\ 0 & 1 \end{pmatrix}$: such a character corresponds to a non-trivial additive character of $\mathbb{F}_q$; there are two classes of such characters under $\mathbf{SL}_2(\mathbb{F}_q)$, which corresponds to the fact that the centre of $\mathbf{SL}_2$ has two connected components (its two elements).

The theory of regular elements and Gelfand-Graev representations is expounded in chapter 14, with, as an application, the computation of all irreducible characters on regular unipotent elements.

Assumed background

We will assume that the reader has some basic familiarity with algebraic geometry, but we will give as far as possible statements of all the results we use; a possible source for these is R. Hartshorne's book "Algebraic Geometry" ([Ha]). Chapters 0, 1 and 2 contain the main results we use from the theory of algebraic groups (the proofs will be often omitted in chapter 0; references for them may be found in the books on algebraic groups by A. Borel [B1], J. E. Humphreys [Hu] and T. A Springer [Sp] which are all good introductions to the subject). We will also recall results about Coxeter groups and root systems for which the most convenient reference is the volume of N. Bourbaki [Bbk] containing chapters IV, V and VI of the theory of Lie groups and Lie algebras. However, we will not give any references for the basic results of the theory of representations of finite groups over fields of characteristic 0, which we assume known. All we need is covered in the first two parts of the book of J. -P. Serre [Se].

Bibliography

Appropriate references will be given for each statement. There are two works about the subject of this book that we will not refer to systematically, but which the reader should consult to get additional material: the book of B. Srinivasan [Sr] for the methods of Deligne and Lusztig, and the survey of R. W. Carter [Ca] which covers many topics that we could not introduce in the span of a one-semester course (such as unipotent classes, Hecke algebras, the work of Kazhdan and Lusztig, ...); furthermore our viewpoint or the organization of our proofs are often quite different from Carter's (for instance the systematic use we make of Mackey's formula (chapters 5 and 11) and of Curtis-Kawanaka-Lusztig duality). To get further references, the reader may look at the quasi-exhaustive bibliography on the subject up to 1986 which is in Carter's book.

Acknowledgments

We would like to thank the "équipe des groupes finis" and the mathematics department of the École Normale Supérieure, who provided us with a stimulating working environment, and adequate facilities for composing this book. We thank also the University of Paris VII which gave us the opportunity to give the course which started this book. We thank all those who read carefully the first drafts and suggested improvements, particularly Michel Enguehard, Guy Rousseau, Jean-Yves Hée, Gunther Malle and the editor. We thank Michel Broué and Jacques Tits who provided us with various ideas and information, and above all George Lusztig who invented most of the theory.

0. BASIC RESULTS ON ALGEBRAIC GROUPS

A finite group of Lie type will be defined in this book as the group of points over a finite field of a (usually connected) reductive algebraic group over the algebraic closure of that finite field. We begin by recalling the definition of these terms and the basic structure theory of reductive algebraic groups.

Let us first establish some notations and conventions we use throughout. If g is an automorphism of a set (resp. variety, group, ...) X, we will denote by X^g the set of fixed points of g, and ${}^g x$ the image of the element $x \in X$ by g. A group G acts naturally on itself by conjugation, and we will hence write ${}^g h$ for ghg^{-1}, where g and h are elements of G. We will write $Z(G)$ for the centre of G; if X is any subset of G, we put $N_G(X) = \{ g \in G \mid {}^g X = X \}$ and $C_G(X) = \{ g \in G \mid gx = xg \text{ for all } x \in X \}$.

We will consider only **affine** algebraic groups over an algebraically closed field k (which will be taken to be $\overline{\mathbb{F}}_q$ from chapter 3 onwards), *i.e.*, affine algebraic varieties endowed with a group structure such that the multiplication and inverse maps are algebraic (which corresponds to a coalgebra structure on the algebra of functions on the variety). For such a group $\mathbf{G}$, we will call elements of $\mathbf{G}$ the elements of the set $\mathbf{G}(k)$ of k-valued points of $\mathbf{G}$. We generally will use bold letters for algebraic groups and varieties.

0.1 Examples of affine algebraic groups.

(i) The **multiplicative group** $\mathbb{G}_m$, and the **additive group** $\mathbb{G}_a$, defined respectively by the algebras $k[T, T^{-1}]$ (with comultiplication $T \mapsto T \otimes T$) and $k[T]$ (with comultiplication $T \mapsto T \otimes 1 + 1 \otimes T$). We have $\mathbb{G}_m(k) \simeq k^\times$ and $\mathbb{G}_a(k) \simeq k^+$.

(ii) The **linear group** $\mathbf{GL}_n$ defined by the algebra $k[T_{i,j}, \det(T_{i,j})^{-1}]$, with comultiplication $T_{i,j} \mapsto \sum_l T_{i,l} \otimes T_{l,j}$; and its subgroup, the **special linear group** $\mathbf{SL}_n$ defined by the ideal generated by $\det(T_{i,j}) - 1$.

Actually every one-dimensional affine algebraic group is isomorphic to $\mathbb{G}_m$ or $\mathbb{G}_a$ (this is surprisingly difficult to prove; see, *e.g.*, [Sp, 2.6]).

A morphism of algebraic groups is a homomorphism of varieties which is also a group homomorphism. A closed subvariety which is a subgroup is naturally an algebraic subgroup (*i.e.*, the inclusion map is a morphism of algebraic groups).

0.2 PROPOSITION. *Let* $\{\mathbf{V}_i\}_{i\in I}$ *be a family of closed subvarieties all containing the identity element of an algebraic group* $\mathbf{G}$*; then the smallest closed subgroup of* $\mathbf{G}$ *containing the* $\mathbf{V}_i$ *is equal to the product* $\mathbf{W}_{i_1}\ldots\mathbf{W}_{i_k}$ *for some finite sequence* $(i_1,\ldots,i_k)$ *of elements of* I*, where either* $\mathbf{W}_i=\mathbf{V}_i$ *or* $\mathbf{W}_i=\{x^{-1}\mid x\in\mathbf{V}_i\}$.

PROOF: See [B1, 2.2]. ■

That subgroup of $\mathbf{G}$ will be called the **subgroup generated** by the $\mathbf{V}_i$.

An algebraic group is called **linear** if it is isomorphic to a closed subgroup of $\mathbb{GL}_n$. It is clear from the definition that a linear algebraic group is affine; the converse is also true (see [Sp, 2.3.5]). Unless otherwise stated, all algebraic groups considered in the sequel will be linear (an example of non-linear algebraic groups are the elliptic curves).

The connected components of an algebraic group $\mathbf{G}$ are finite in number and coincide with its irreducible components; the component containing the identity element of $\mathbf{G}$ is called the **identity component** and denoted by $\mathbf{G}^\circ$. It is a characteristic subgroup of $\mathbf{G}$, and the quotient $\mathbf{G}/\mathbf{G}^\circ$ is finite. Conversely every normal closed subgroup of finite index contains $\mathbf{G}^\circ$ (these properties are elementary; see, *e.g.*, [Sp, 2.2.1]). It is clear from 0.2 that the subgroup generated by a family of connected subvarieties is connected.

0.3 JORDAN DECOMPOSITION.

An element of a linear algebraic group $\mathbf{G}$ is called **semi-simple** (resp. **unipotent**) if its image in some embedding of $\mathbf{G}$ in a $\mathbb{GL}_n$ is semi-simple (resp. unipotent); this property does not depend on the embedding. Every element has a unique decomposition (its **Jordan decomposition**) as the product of two commuting semi-simple and unipotent elements (see [Sp, 2.4.8]).

Tori, solvable groups, Borel subgroups

0.4 DEFINITION.

(i) A **torus** *is an algebraic group which is isomorphic to the product of a finite number of copies of the multiplicative group.*

(ii) A **rational character** *of an algebraic group* $\mathbf{G}$ *is an algebraic group morphism from* $\mathbf{G}$ *to* $\mathbb{G}_m$.

(iii) *The* **character group** $X(\mathbf{T})$ *of a torus* $\mathbf{T}$ *is the group of rational characters of* $\mathbf{T}$.

If $\mathbf{T}$ is a torus isomorphic to $\mathbb{G}_m^r$, its algebra is isomorphic to $k[t_1,\ldots,t_r,t_1^{-1},\ldots,t_r^{-1}]$. An element of $X(\mathbf{T})$ corresponds to a morphism of both algebras and coalgebras $k[t,t^{-1}] \to k[t_1,\ldots,t_r,t_1^{-1},\ldots,t_r^{-1}]$, which is defined by giving the image of t which must be an invertible element, *i.e.*, a monomial; this monomial must be unitary for the morphism to be a coalgebra morphism. The multiplication (multiplying the values) of characters of $\mathbf{T}$ corresponds to the multiplication of monomials; so that for that multiplication $X(\mathbf{T})$ is isomorphic to the group of unitary monomials, and so the algebra of the torus is isomorphic to the group algebra $k[X(\mathbf{T})]$ of $X(\mathbf{T})$ over k.

Let us recall some results about tori.

0.5 PROPOSITION. *Given a torus* $\mathbf{T}$ *and a subtorus* $\mathbf{S}$ *of* $\mathbf{T}$, *there exists a subtorus* $\mathbf{S}'$ *of* $\mathbf{T}$ *such that* $\mathbf{T}=\mathbf{S}\mathbf{S}'$ *(direct product).*

PROOF: The inclusion morphism from the closed subgroup $\mathbf{S}$ into $\mathbf{T}$ corresponds to a surjective morphism of the algebra of the variety $\mathbf{T}$ onto that of $\mathbf{S}$. As explained above, these algebras are respectively isomorphic to $k[X(\mathbf{T})]$ and $k[X(\mathbf{S})]$; the morphism sends $X(\mathbf{T})$ to $X(\mathbf{S})$ by sending a character of $\mathbf{T}$ to its restriction to $\mathbf{S}$. Its surjectivity implies that the morphism comes from a surjective morphism of free abelian groups $X(\mathbf{T}) \to X(\mathbf{S})$, *i.e.*, that every character of $\mathbf{S}$ extends to $\mathbf{T}$. This morphism is consequently split, *i.e.*, $X(\mathbf{S})$ can be identified with a sublattice of $X(\mathbf{T})$ and the kernel X' of the morphism is a supplementary sublattice to $X(\mathbf{S})$. So the closed subgroup of $\mathbf{T}$ whose algebra is $k[X']$ is a torus $\mathbf{S}'$ such that $\mathbf{T}=\mathbf{S}\mathbf{S}'$. ■

0.6 PROPOSITION. *A closed connected subgroup of a torus is a torus, as well as the quotient of a torus by a closed subgroup .*

PROOF: See [Sp, 2.5.2, 2.5.3, 2.5.8 (ii)]. ■

0.7 PROPOSITION. *Given an algebraic action of a torus* $\mathbf{T}$ *on an affine variety* $\mathbf{X}$, *there exists an element* $t\in\mathbf{T}$ *such that* $\mathbf{X}^t=\mathbf{X}^{\mathbf{T}}$.

PROOF: It is known that the action can be linearized, *i.e.*, there exists an embedding of $\mathbf{X}$ into a finite-dimensional vector space V and an embedding of $\mathbf{T}$ into $\mathbf{GL}(V)$ such that the action of $\mathbf{T}$ factors through this embedding (see [P. SLODOWY Simple Singularities and Simple Algebraic Groups, *Lecture notes in mathematics,* **815** (1980), Springer, I, 1.3]). The space V has a $\mathbf{T}$-stable decomposition indexed by characters in $X(\mathbf{T})$. The kernel of some non-trivial $\chi \in X(\mathbf{T})$ is of codimension 1 in $\mathbf{T}$ since its algebra is the quotient of the algebra of $\mathbf{T}$ by the ideal generated by χ (identifying the algebra of $\mathbf{T}$ with the group algebra of $X(\mathbf{T})$). It follows that there is some element $t \in \mathbf{T}$ which lies outside the kernels of all non-trivial characters of $\mathbf{T}$ occuring in V; the fixed points of t in V (and so in $\mathbf{X}$) are the same as those of $\mathbf{T}$ whence the result. ∎

0.8 THEOREM (rigidity of tori). *Let* $\mathbf{T}$ *be a torus of an algebraic group* $\mathbf{G}$, *then* $N_{\mathbf{G}}(\mathbf{T})^{\circ} = C_{\mathbf{G}}(\mathbf{T})^{\circ}$.

PROOF: See, *e.g.*, [Sp, 2.5.11]. ∎

It follows in particular that the quotient $N_{\mathbf{G}}(\mathbf{T})/C_{\mathbf{G}}(\mathbf{T})$ is finite.

0.9 PROPOSITION. *For a solvable algebraic group* $\mathbf{G}$:

(i) Every semi-simple element of $\mathbf{G}^{\circ}$ *lies in a maximal torus of* $\mathbf{G}$.
(ii) All maximal tori of $\mathbf{G}$ *are conjugate.*
(iii) If $\mathbf{G}$ *is connected, the set* $\mathbf{G}_u$ *of unipotent elements of* $\mathbf{G}$ *is a normal connected subgroup, and for every maximal torus* $\mathbf{T}$ *of* $\mathbf{G}$, *there is a semi-direct product decomposition* $\mathbf{G} = \mathbf{G}_u \rtimes \mathbf{T}$.

PROOF: See, *e.g.*, [Sp, 6.11] (the assertions on $\mathbf{G}_u$ follow from the theorem of Lie-Kolchin, which states that every closed solvable subgroup of $\mathbf{GL}_n(k)$ is conjugate to a subgroup of the group of upper triangular matrices). ∎

Let us note that this proposition implies that every connected solvable algebraic group containing no unipotent elements is a torus.

0.10 DEFINITION. *Maximal closed connected solvable subgroups of an algebraic group are called* **Borel subgroups**.

These groups are of paramount importance in the theory. The next theorem states their basic properties.

0.11 THEOREM. *Let* $\mathbf{G}$ *be a connected algebraic group. Then:*

(i) All Borel subgroups of $\mathbf{G}$ *are conjugate.*
(ii) Every element of $\mathbf{G}$ *is in some Borel subgroup.*
(iii) The centralizer in $\mathbf{G}$ *of any torus is connected.*
(iv) A Borel subgroup is equal to its normalizer in $\mathbf{G}$.

PROOF: For a detailed proof see [Sp, 7.2.6, 7.3.3, 7.3.5 and 7.3.7]. Here is an outline. To show (i), one first shows that $\mathbf{G}/\mathbf{B}$ is a complete variety; a complete variety has the property that a connected solvable group acting on it always has a fixed point. Thus another Borel subgroup $\mathbf{B}'$ acting on $\mathbf{G}/\mathbf{B}$ by left translation has a fixed point, *i.e.*, there exists $g \in \mathbf{G}$ such that $\mathbf{B}'g\mathbf{B} = g\mathbf{B}$, so $g^{-1}\mathbf{B}'g \subset \mathbf{B}$ whence the result. Property (ii) similarly results from properties of complete varieties.

To show (iii) one uses (ii) and properties of $\mathbf{G}/\mathbf{B}$ to prove that if x centralizes some torus $\mathbf{S}$, there is some Borel subgroup containing both x and $\mathbf{S}$. Thus it is sufficient to show that (iii) holds for a solvable group, which follows from 0.9 (iii). Property (iv) is proved by induction on the dimension of $\mathbf{G}$, using (iii) to get a connected group of dimension less than that of $\mathbf{G}$. ■

It follows from (iv) above and the remark that the closure of a solvable group is solvable (see [B1, 2.4]) that the words "closed connected" can be omitted from the definition of a Borel subgroup.

0.12 COROLLARY. *Let $\mathbf{G}$ be a connected algebraic group.*

(i) *Any closed subgroup containing a Borel subgroup is equal to its normalizer in $\mathbf{G}$ and is connected.*
(ii) *Two closed subgroups containing the same Borel subgroup and conjugate in $\mathbf{G}$ are equal.*
(iii) *All maximal tori of $\mathbf{G}$ are conjugate; every semi-simple element of $\mathbf{G}$ is in some maximal torus.*
(iv) *Two elements of a maximal torus $\mathbf{T}$ conjugate under $\mathbf{G}$ are conjugate under $N_{\mathbf{G}}(\mathbf{T})$.*

PROOF: Property (i) results from 0.11 (i) and 0.11 (iv): if $\mathbf{P} \supset \mathbf{B}$, and g normalizes $\mathbf{P}$, then g normalizes $\mathbf{P}^\circ$. Thus $\mathbf{B}$ and ${}^g\mathbf{B}$ are Borel subgroups of $\mathbf{P}^\circ$, so are conjugate by some $p \in \mathbf{P}^\circ$, so that ${}^{pg}\mathbf{B} = \mathbf{B}$; by 0.11 (iv) it follows that $pg \in \mathbf{B}$, whence $g \in \mathbf{P}^\circ$. We have proved that $N_{\mathbf{G}}(\mathbf{P}) \subset \mathbf{P}^\circ$, which proves (i). The proof of (ii) is similar (see [Sp, 7.3.9]). Since any maximal torus is in some Borel subgroup, the first assertion in (iii) results from 0.11 (i) and 0.9 (ii). The second assertion of (iii) comes from 0.11 (ii) and 0.9 (i).

Let us prove (iv). If s and ${}^g s$ both lie in $\mathbf{T}$ then $\mathbf{T}$ and ${}^{g^{-1}}\mathbf{T}$ are two maximal tori containing s, so are two maximal tori of the group $C_{\mathbf{G}}(s)$, and since they are connected they lie in the identity component $C_{\mathbf{G}}(s)^\circ$. By (iii) they are conjugate by some element $x \in C_{\mathbf{G}}(s)^\circ$, *i.e.*, ${}^x\mathbf{T} = {}^{g^{-1}}\mathbf{T}$ so $gx \in N_{\mathbf{G}}(\mathbf{T})$. As ${}^{gx}s = {}^g s$, we get the result. ■

0.13 DEFINITION. *A* **unipotent** *algebraic group is a group containing only unipotent elements.*

0.14 PROPOSITION. *Every unipotent subgroup of an affine algebraic group is nilpotent.*

PROOF: This property is invariant by an embedding into $\mathbf{GL}_n$, and any unipotent subgroup of $\mathbf{GL}_n(k)$ is conjugate to a subgroup of the group of upper triangular matrices which have all their diagonal coefficients equal to 1 (see [Sp, 2.4.11] for a proof), whence the result. ■

0.15 COROLLARY. *The maximal connected unipotent subgroups of an algebraic group* $\mathbf{G}$ *are the groups* $\mathbf{B}_u$ *for* $\mathbf{B}$ *in the set of Borel subgroups of* $\mathbf{G}$.

PROOF: The proof of 0.14 shows that such a subgroup is in a Borel subgroup whence the result by 0.9 (iii). ■

Radical, unipotent radical, reductive and semi-simple groups

0.16 PROPOSITION. *Let* $\mathbf{G}$ *be an algebraic group.*

(i) The product of all the closed connected normal solvable subgroups of $\mathbf{G}$ *is also a closed connected normal solvable subgroup of* $\mathbf{G}$ *called the* **radical** *of* $\mathbf{G}$ *and denoted by* $R(\mathbf{G})$.

(ii) Similarly the set of all closed connected normal unipotent subgroups of $\mathbf{G}$ *has a unique maximal element called the* **unipotent radical** *of* $\mathbf{G}$ *and denoted by* $R_u(\mathbf{G})$.

(iii) $R_u(\mathbf{G}) = R(\mathbf{G})_u$ *(where* $R(\mathbf{G})_u$ *is defined as in 0.9 (iii)).*

PROOF: Using 0.2 it follows that the product in (i) is actually finite. This implies (i) since the product of two solvable groups normalizing each other is still solvable. To see (ii) and (iii) we first remark that a closed connected normal unipotent subgroup is in $R(\mathbf{G})$, so in $R(\mathbf{G})_u$. We then observe that $R(\mathbf{G})_u$ is normal in $\mathbf{G}$, being characteristic in $R(\mathbf{G})$, and is connected by 0.9 (iii). ■

0.17 DEFINITION. *An algebraic group is called* **reductive** *if its unipotent radical is trivial, and* **semi-simple** *if its radical is trivial.*

0.18 PROPOSITION. *If* $\mathbf{G}$ *is connected and reductive, then* $R(\mathbf{G}) = Z(\mathbf{G})^\circ$, *the identity component of the centre of* $\mathbf{G}$.

PROOF: By 0.9 (iii), since $R_u(\mathbf{G})$ is trivial, $R(\mathbf{G})$ is a torus normal in $\mathbf{G}$. Since $\mathbf{G}$ is connected and for any torus $\mathbf{T}$ we have $N_{\mathbf{G}}(\mathbf{T})^\circ = C_{\mathbf{G}}(\mathbf{T})^\circ$ (see 0.8), $R(\mathbf{G})$ is central in $\mathbf{G}$, and, being connected, is in $Z(\mathbf{G})^\circ$. Conversely $Z(\mathbf{G})^\circ$, being normal solvable and connected, is contained in $R(\mathbf{G})$. ■

0.19 PROPOSITION. *If* $\mathbf{G}$ *is reductive and connected, its derived group is semi-simple and has a finite intersection with* $Z(\mathbf{G})$.

PROOF: The first assertion results from the second one since the radical $R(\mathbf{G}')$ of the derived group of $\mathbf{G}$ being characteristic (by 0.18) is in $R(\mathbf{G})$, so in $Z(\mathbf{G}) \cap \mathbf{G}'$; it is trivial since it is connected and this last group is finite.

To see the second assertion we may imbed $\mathbf{G}$ in some $\mathbf{GL}(V)$; the space V is a direct sum of isotypic spaces V_χ where χ runs over the rational characters of $Z(\mathbf{G})$. The action of $\mathbf{G}$ preserves this decomposition, so the image of $\mathbf{G}$ is in $\prod \mathbf{GL}(V_\chi)$ and that of $\mathbf{G}'$ in $\prod \mathbf{SL}(V_\chi)$ while that of $Z(\mathbf{G})$ consists of products of scalar matrices in each V_χ, whence the result. ■

EXAMPLES. The group $\mathbf{GL}_n$ is reductive. The group $\mathbf{SL}_n$ is semi-simple as it has a finite centre.

One-parameter subgroups, roots, coroots, structure theorem for reductive groups

Let $\mathbf{T}$ be a torus. Algebraic group homomorphisms from $\mathbf{G}_m$ to $\mathbf{T}$ are called **one-parameter subgroups** of $\mathbf{T}$; they form an abelian group denoted by $Y(\mathbf{T})$. An element of $Y(\mathbf{T})$ corresponds to a morphism of algebras $k[t_1, \ldots, t_r, t_1^{-1}, \ldots, t_r^{-1}] \to k[t, t^{-1}]$, which is determined by the images of the t_i which must be invertible elements, *i.e.*, monomials. The multiplication of one-parameter subgroups corresponds to the multiplication of monomials so $Y(\mathbf{T})$ is isomorphic to a product of r copies of the group of monomials in $k[t, t^{-1}]$.

There is an exact pairing between $X(\mathbf{T})$ and $Y(\mathbf{T})$ (*i.e.*, a map $X(\mathbf{T}) \times Y(\mathbf{T}) \to \mathbf{Z}$ which makes each one the $\mathbf{Z}$-dual of the other) obtained as follows: given $\chi \in X(\mathbf{T})$ and $\psi \in Y(\mathbf{T})$ the composite map $\chi \circ \psi$ is a homomorphism from $\mathbf{G}_m$ to itself, so is of the form $x \mapsto x^n$ for some $n \in \mathbf{Z}$; the map $X(\mathbf{T}) \times Y(\mathbf{T}) \to \mathbf{Z}$ is defined by $(\chi, \psi) \mapsto n$.

0.20 PROPOSITION. *The map $y \otimes x \mapsto y(x)$ is a group isomorphism: $Y(\mathbf{T}) \otimes_{\mathbf{Z}} k^\times \xrightarrow{\sim} \mathbf{T}$.*

PROOF: Let $(x_i)_{i=1,\ldots,n}$ and $(y_i)_{i=1,\ldots,n}$ be two dual bases of $X(\mathbf{T})$ and $Y(\mathbf{T})$ respectively. It is easy to check that $t \mapsto \sum_{i=1}^{i=n} y_i \otimes x_i(t) : \mathbf{T} \to Y(\mathbf{T}) \otimes_{\mathbf{Z}} k^\times$ is the inverse of the map of the statement, using the fact that $\bigcap_{x \in X(\mathbf{T})} \ker x = 1$ (this follows from the isomorphism between the algebra of the variety $\mathbf{T}$ and $k[X(\mathbf{T})]$; see remarks after 0.4). ■

We will need the following property of X seen as a functor from tori to $\mathbf{Z}$-modules:

0.21 PROPOSITION. *The functor X is exact.*

PROOF: This is an immediate corollary of the fact that a subtorus is always a direct factor (see 0.5). ■

We now study the relationship between closed subgroups of $\mathbf{T}$ and subgroups of $X(\mathbf{T})$.

0.22 LEMMA. *Given a torus* $\mathbf{T}$, *let* $x_1, \ldots, x_n$ *be linearly independent elements of* $X(\mathbf{T})$ *and* $\lambda_1, \ldots, \lambda_n$ *arbitrary elements of* $k^\times$; *then there exists* $s \in \mathbf{T}$ *such that* $x_i(s) = \lambda_i$ *for* $i = 1, \ldots, n$.

PROOF: See, *e.g.*, [Hu, 16.2, lemma C]. ■

0.23 DEFINITION. *Given a torus* $\mathbf{T}$ *and a closed subgroup* $\mathbf{S}$ *of* $\mathbf{T}$, *we define* $\mathbf{S}^\perp = \{ x \in X(\mathbf{T}) \mid \forall s \in \mathbf{S}, \quad x(s) = 1 \}$; *and conversely, given a subgroup* A *of* $X(\mathbf{T})$, *we define a subgroup of* $\mathbf{T}$ *by* $A^\perp = \{ s \in \mathbf{T} \mid \forall x \in A, \quad x(s) = 1 \}$ *(which is closed since* A *is finitely generated).*

0.24 PROPOSITION. *If* k *is of characteristic* p, *given a torus* $\mathbf{T}$ *and a subgroup* A *of* $X(\mathbf{T})$, *the group* $A^{\perp\perp}/A$ *is the* p*-torsion subgroup of* $X(\mathbf{T})/A$.

PROOF: First notice that, for any closed subgroup $\mathbf{S}$ of $\mathbf{T}$, the group $X(\mathbf{T})/\mathbf{S}^\perp$ has no p-torsion. Indeed,

$$p^n x \in \mathbf{S}^\perp \Leftrightarrow \forall s \in \mathbf{S}, \quad x(s)^{p^n} = 1 \Leftrightarrow \forall s \in \mathbf{S}, \quad x(s) = 1 \Leftrightarrow x \in \mathbf{S}^\perp,$$

where the middle equivalence holds since $x \mapsto x^{p^n}$ is an automorphism of k.

Thus it is enough to see that $A^{\perp\perp}/A$ is a p-group. Let $x \in A^{\perp\perp} - A$. It is a standard result on free $\mathbb{Z}$-modules that there is a basis $(x, x_1, \ldots, x_r)$ of $<A, x>$ such that $(mx, x_1, \ldots, x_r)$ is a basis of A (with $m \in \mathbb{Z}$; possibly $m = 0$). The result clearly follows if we can prove that m is a power of p. Let us assume otherwise: then there exists $\lambda \in k^\times$ such that $\lambda \neq 1$ and $\lambda^m = 1$ (even if $m = 0$). By 0.22, there exists some $s \in \mathbf{T}$ such that $x(s) = \lambda, \quad x_1(s) = 1, \ldots, \quad x_r(s) = 1$. Thus $mx(s) = 1$ so $s \in A^\perp$ but $x(s) \neq 1$, which contradicts $x \in A^{\perp\perp}$. ■

We will now give the main structure theorems for reductive groups. In order to do that, we first recall the definition and some properties of root systems (the reader can look at the first two items of 0.31 to see why they appear in this context).

0.25 DEFINITION. *A* **root system** *in a real vector space* V *is a subset* Φ *with the following properties*

(i) Φ *is finite, generates* V *and* $0 \notin \Phi$.

(ii) *For any* $\alpha \in \Phi$, *there exists* $\check{\alpha}$ *in the vector space dual to* V *such that* $\langle \alpha, \check{\alpha} \rangle = 2$ *and such that* Φ *is stable under the reflection* $s_\alpha : V \to V$ *defined by* $x \mapsto x - \langle x, \check{\alpha} \rangle \alpha$.

(iii) $\check{\alpha}(\Phi) \subset \mathbb{Z}$ *for any* $\alpha \in \Phi$.

Then the $\check{\alpha}$ form a root system $\check{\Phi}$ in the dual of V. They are called **coroots**; there exists a scalar product on V invariant by the s_α, which lets us identify V with its dual (see [Bbk, VI, 1.1, propositions 1 and 2]). Under this identification $\check{\alpha}$ becomes $2\alpha/\langle \alpha, \alpha \rangle$. Let us note that by (ii), if α is a root, then $-\alpha$ is another one. The root system is said to be **reduced** if any line in V containing a root contains exactly two (opposed) roots. Let us note also that if Φ is the union of two orthogonal subsets then each of them is a root system in the subspace of V that it generates. A root system is **irreducible** if there exists no such decomposition.

The group W generated by the s_α's is called the **Weyl group** of the root system; it is a group of permutations of Φ by 0.25 (i) and (ii). We will give a presentation of it which shows that it is a Coxeter group. In order to do that we first introduce bases of root systems.

0.26 DEFINITION. *Given an ordered vector space structure on* V *such that any root is positive or negative, we denote by* Φ^+ *(resp.* Φ^-*) the set of positive (resp. negative) roots. Positive roots which are indecomposable into a sum of other positive roots are called* **simple roots**. *The set of simple roots is called the* **basis** *of* Φ *relative to the given order.*

0.27 PROPOSITION. *A subset* $\Pi \subset \Phi$ *is a basis for some order if and only if* Π *is a basis of* V *and every element of* Φ *is a linear combination of elements of* Π *with integral coefficients which all have the same sign.*

PROOF: See [Bbk, VI, 1.6 theorem 3 and 1.7 corollary 3 of proposition 20]. ■

It follows that to be given a basis of Φ and to be given the set Φ^+ are equivalent.

0.28 PROPOSITION. *Let* Π *be a basis of* Φ*; then*

$$<\{s_\alpha\}_{\alpha\in\Pi}; {s_\alpha}^2 = 1, (s_\alpha s_\beta)^{m_{\alpha,\beta}} = 1>$$

is a presentation of the Weyl group of Φ, *where* $m_{\alpha,\beta}$ *is the order (which is finite) of the product* $s_\alpha s_\beta$.

PROOF: See [Bbk, VI, 1.5 theorem 2]. ■

This gives a presentation of W as a **Coxeter group** (a group with a set of generators S having a presentation like 0.28). In a Coxeter group W, the **length** $l(w)$ of $w \in W$ is the minimum number of generators in an expression of the element (such a minimal expression is called **reduced**). The length l is relative to the generating set S, *i.e.*, in the Weyl group of a root system it depends on the choice of a basis Π; it can actually be expressed in another way, namely:

0.29 PROPOSITION. *The length of $w \in W$ is equal to the number of roots in Φ^+ which are sent to Φ^- under the action of w.*

PROOF: See [Bbk, VI, 1.6 corollary 2 of proposition 17]. ■

One of the basic properties of Coxeter groups is the "exchange lemma":

0.30 LEMMA. *Let W be a Coxeter group, let $w \in W$ and s be a generator such that $l(ws) < l(w) + 1$. Let $w = s_1 s_2 \ldots s_{l(w)}$ be a reduced expression for w; then there exists $i \in \{1, \ldots, l(w)\}$ such that $ws = s_1 \ldots \hat{s}_i \ldots s_{l(w)}$ (a product from which the term s_i has been omitted).*

PROOF: See [Bbk, IV, 1.5 proposition 4]. ■

EXERCISE. Prove this result for the Weyl group of a root system using just the characterization of the length given by 0.29.

In the following theorem, which describes the key facts for the study of the structure of reductive groups, $\mathbf{G}$ is a connected reductive group and $\mathbf{T}$ a maximal torus of $\mathbf{G}$.

0.31 THEOREM.

(i) *Non-trivial minimal closed unipotent subgroups of $\mathbf{G}$ normalized by $\mathbf{T}$ are isomorphic to $\mathbb{G}_a$; the conjugation action of $\mathbf{T}$ is mapped by this isomorphism to an action of $\mathbf{T}$ on $\mathbb{G}_a$ of the form $x \mapsto \alpha(t)x$, where $\alpha \in X(\mathbf{T})$.*

(ii) *The elements $\alpha \in X(\mathbf{T})$ thus obtained are all distinct and non-zero and are finite in number. They form a (reduced) root system Φ in the subspace of $X(\mathbf{T}) \otimes \mathbb{R}$ that they generate (they are called the* **roots of G relative to T**). *The group $W(\mathbf{T}) = N_{\mathbf{G}}(\mathbf{T})/\mathbf{T}$ is isomorphic to the Weyl group of Φ.*

(iii) *The group $\mathbf{G}$ is generated by $\mathbf{T}$ and $\{\mathbf{U}_\alpha\}_{\alpha \in \Phi}$ where $\mathbf{U}_\alpha$ is the unipotent subgroup corresponding to α by (i) and (ii).*

(iv) *If $\alpha \neq -\beta$ then the set of commutators $[\mathbf{U}_\alpha, \mathbf{U}_\beta]$ is contained in $\prod_\gamma \mathbf{U}_\gamma$ where γ runs over the set, taken in any order, of roots of the form $a\alpha + b\beta$ with a and b positive integers.*

(v) *The Borel subgroups containing $\mathbf{T}$ correspond one-to-one to bases of Φ; if Φ^+ is the set of positive roots corresponding to such a basis, the* *corresponding Borel subgroup is equal to* $\mathbf{T}\prod_{\alpha\in\Phi^+}\mathbf{U}_\alpha$ *for any order on* Φ^+.

PROOF: Roots are usually introduced using the Lie algebra of **G**. A group-theoretic approach is possible using (i). We just note here that by 0.33 below (which is proved in [Hu] with the Lie algebra definition of the roots) (i) is equivalent to other definitions in the literature. It follows then from, *e.g.*, [Sp, 9.3.6 (see also 9.3.9 (i))] that (i)–(iii) hold. For (iv) see [Sp, 10.1.4] and for (v) see [Sp, 10.1.1 and 10.1.5]. ∎

We will sometimes write W for $W(\mathbf{T})$ when there is no ambiguity for **T**. This group is called the **Weyl group** of **G** with respect to **T**. Let us note that by 0.12 (iii) the group $W(\mathbf{T})$ and Φ do not depend on **T** up to isomorphism.

The next proposition is used in the proof of 0.31.

0.32 PROPOSITION.

(i) *The unipotent radical of an algebraic group* **G**, *is equal to the unipotent radical of the intersection of the Borel subgroups containing a given maximal torus.*

(ii) *In a connected reductive group, the centralizer of any torus is reductive (and connected by 0.11(iii)).*

(iii) *In a connected reductive group, every maximal torus is equal to its centralizer.*

PROOF: See [Sp, 9.3.4 and 9.3.5]. ∎

Until the end of this chapter the notation is as in 0.31.

0.33 PROPOSITION. *Every closed unipotent subgroup* **V** *of* **G** *normalized by* **T** *is connected and equal to the product* $\prod_{\mathbf{U}_\alpha\subset\mathbf{V}}\mathbf{U}_\alpha$ *for any order on the* α*'s.*

PROOF: From the remark following 0.11 the group **V** is in some Borel subgroup; the result in that case can be found in [Hu, proposition 28.1]. ∎

0.34 PROPOSITION. *Every closed connected subgroup* **H** *of* **G** *normalized by* **T** *is generated by* $(\mathbf{T}\cap\mathbf{H})^\circ$ *and the* $\mathbf{U}_\alpha$ *it contains.*

PROOF: see [BT, 3.4]. ∎

0.35 PROPOSITION. *The centre* $Z(\mathbf{G})$ *is the intersection in* **T** *of the kernels of the roots relative to* **T**.

PROOF: By 0.32 we have $Z(\mathbf{G}) \subset C_{\mathbf{G}}(\mathbf{T}) = \mathbf{T}$. An element of $\mathbf{T}$ central in $\mathbf{G}$ must act trivially on all $\mathbf{U}_\alpha$, so be in the kernel of all the roots. Conversely, since (see 0.31 (iii)) the $\mathbf{U}_\alpha$ and $\mathbf{T}$ generate $\mathbf{G}$, such an element centralizes $\mathbf{G}$. ■

0.36 COROLLARY. *The centre of* $\mathbf{G}/Z(\mathbf{G})$ *is trivial.*

PROOF: The image in $\mathbf{G}/Z(\mathbf{G})$ of an element $t \in \mathbf{T}$ is central if and only if the commutator of t with any element of $\mathbf{G}$ is in $Z(\mathbf{G})$. But the commutator of t with an element of $\mathbf{U}_\alpha$ is in $\mathbf{U}_\alpha$, so it has to be 1 as $\mathbf{U}_\alpha \cap Z(\mathbf{G}) = \{1\}$. So t is in the kernel of all the roots, which is $Z(\mathbf{G})$ by 0.35. ■

0.37 PROPOSITION. *Let* $\mathbf{G}$ *be a connected reductive group. Then each of the following properties is equivalent to* $\mathbf{G}$ *being semi-simple*

(i) $\mathbf{G}$ *is generated by the* $\mathbf{U}_\alpha$.

(ii) $\mathbf{G}$ *is equal to its derived group.*

(iii) For any maximal torus $\mathbf{T}$, *the roots of* $\mathbf{G}$ *with respect to* $\mathbf{T}$ *generate* $X(\mathbf{T}) \otimes \mathbb{R}$.

PROOF: For the fact that a semi-simple group satisfies (i)–(iii) see, *e.g.*, [Sp, 9.4.1 and 9.4.2]. It is also shown there that in a reductive group $\mathbf{G}$ we have $\mathbf{U}_\alpha \subset \mathbf{G}'$. Thus a group satisfying (i) satisfies (ii); such a group is clearly semi-simple by 0.19. Finally if (iii) holds then $X(\mathbf{T})/<\Phi>$ is finite where $<\Phi>$ is the sublattice of $X(\mathbf{T})$ spanned by Φ. Since X is exact and Φ is in the kernel of the restriction map $X(\mathbf{T}) \to X(Z(\mathbf{G})^\circ)$ (see 0.35) it follows that $X(Z(\mathbf{G})^\circ)$ is finite (so trivial) and so $R(\mathbf{G}) = Z(\mathbf{G})^\circ$ (see 0.18) is trivial. ■

0.38 PROPOSITION. *A connected semi-simple group* $\mathbf{G}$ *has finitely many minimal non-trivial normal connected subgroups. Any two of them commute and each of them has a finite intersection with the product of the others; the product of all of them is equal to the whole of* $\mathbf{G}$.

PROOF: See again [Sp, 9.4.1]. ■

0.39 DEFINITION. *A connected semi-simple group which has no non-trivial proper normal connected subgroup is called* **quasi-simple**.

0.40 PROPOSITION. *A connected reductive group is the product of its derived group by its radical.*

PROOF: It is equivalent to see that the composite morphism $\mathbf{G}' \hookrightarrow \mathbf{G} \to \mathbf{G}/R\mathbf{G}$ is surjective. But $\mathbf{G}'/(R\mathbf{G} \cap \mathbf{G}') = (\mathbf{G}/R\mathbf{G})' = \mathbf{G}/R\mathbf{G}$, this last equality by 0.37 (ii) since $\mathbf{G}/R\mathbf{G}$ is semi-simple. ■

The above product is almost direct (*i.e.*, the intersection is finite) by 0.19.

0.41 COROLLARY. *Every connected reductive group is the almost direct product of its radical and of a finite number of quasi-simple groups.*

PROOF: This is just 0.38 and 0.40 put together. ∎

We will use the following notation.

0.42 NOTATION. *Suppose we are given two algebraic varieties* $\mathbf{X}$ *and* $\mathbf{Y}$*, and a right action on* $\mathbf{X}$ *and a left action on* $\mathbf{Y}$ *of the finite group* G*; we will denote by* $\mathbf{X} \times_G \mathbf{Y}$ *the quotient of* $\mathbf{X} \times \mathbf{Y}$ *by the diagonal (left) action of* G *where* $g \in G$ *acts by* (g^{-1}, g).

0.43 PROPOSITION. *A connected reductive group is quasi-simple if and only if its root system is irreducible.*

PROOF: Using 0.31 (iv) it is easy to see that a decomposition of Φ into two orthogonal root systems corresponds to a decomposition of $\mathbf{G}$ as a product of two normal connected subgroups.

Conversely, we may assume that $\mathbf{G} = \mathbf{G}_1.\mathbf{G}_2$, an almost direct product, *i.e.*, $\mathbf{G} \simeq \mathbf{G}_1 \times_Z \mathbf{G}_2$ where $Z = \mathbf{G}_1 \cap \mathbf{G}_2$ is a finite central subgroup. Using that isomorphism and the isomorphism $\mathbf{G}/Z \simeq \mathbf{G}_1/Z \times \mathbf{G}_2/Z$, it is easy to see that maximal tori of $\mathbf{G}$ are of the form $\mathbf{T}_1 \times_Z \mathbf{T}_2$ where $\mathbf{T}_1$ (resp. $\mathbf{T}_2$) is a maximal torus of $\mathbf{G}_1$ (resp. $\mathbf{G}_2$), and that the roots of $\mathbf{G}$ relative to $\mathbf{T}$ are the disjoint union of those of $\mathbf{G}_1$ relative to $\mathbf{T}_1$ and of those of $\mathbf{G}_2$ relative to $\mathbf{T}_2$, whence the result. ∎

We recall that irreducible root systems are classified in four infinite series A_n, B_n, C_n, D_n and five "exceptional" systems E_6, E_7, E_8, F_4 and G_2 (see [Bbk, VI, 4.2, theorem 3]).

0.44 PROPOSITION.

(i) *For any* $\alpha \in \Phi$*, there exists a homomorphism* $\mathbf{SL}_2 \to \mathbf{G}$ *whose image is the group* $\langle \mathbf{U}_\alpha, \mathbf{U}_{-\alpha} \rangle$*, and such that the image of the subgroup of matrices of the form* $\begin{pmatrix} 1 & * \\ 0 & 1 \end{pmatrix}$ *(resp.* $\begin{pmatrix} 1 & 0 \\ * & 1 \end{pmatrix}$*) is* $\mathbf{U}_\alpha$ *(resp.* $\mathbf{U}_{-\alpha}$*). This homomorphism is unique up to composition with conjugacy by an element of the form* $\begin{pmatrix} a & 0 \\ 0 & a^{-1} \end{pmatrix} \in \mathbf{SL}_2$.

(ii) *The one-parameter subgroup of* $\mathbf{T}$ *which sends* $x \in \mathbb{G}_m$ *to the image of the matrix* $\begin{pmatrix} x & 0 \\ 0 & x^{-1} \end{pmatrix}$ *by the above homomorphism is identified via the pairing between* $Y(\mathbf{T})$ *and* $X(\mathbf{T})$ *with a linear form on* $X(\mathbf{T})$ *whose restriction to* $\langle \Phi \rangle$ *is the coroot* $\check{\alpha}$.

(iii) *The element of $W(\mathbf{T})$ (which we will denote again by s_α) corresponding by the isomorphism of 0.31 (ii) to the element s_α in the Weyl group of Φ is the image of* $\begin{pmatrix} 0 & 1 \\ -1 & 0 \end{pmatrix}$ *by the above homomorphism and the quotient by* $\mathbf{T}$.

PROOF: See, *e.g.*, [Sp, 11.2.1]. ■

Note that this proposition implies that $<\mathbf{U}_\alpha, \mathbf{U}_{-\alpha}>$ is isomorphic to $\mathbf{SL}_2$ or $\mathbf{PGL}_2$ (the root system of $\mathbf{SL}_2$ is of type A_1). Note also that from (iii) above for $\alpha \in \Phi$ we have $\beta \circ \operatorname{ad} s_\alpha = s_\alpha(\beta)$ where s_α stands in the left-hand side for the element of $W(\mathbf{T})$ and in the right-hand side for the reflection in $X(\mathbf{T}) \otimes \mathbb{R}$, whence $s_\alpha \mathbf{U}_\beta {s_\alpha}^{-1} = \mathbf{U}_{s_\alpha(\beta)}$.

0.45 THEOREM (CLASSIFICATION). *The datum $(X(\mathbf{T}), Y(\mathbf{T}), \Phi, \check{\Phi})$ characterizes* **G** *up to isomorphism (it is called the* **root datum** *of* **G***); and each possible root datum is the root datum of some reductive group.*

PROOF: See [Sp, 11.4.3 and 12.1] or [Hu, 32.1]. ■

1. THE BRUHAT DECOMPOSITION; PARABOLIC SUBGROUPS.

In this chapter we review properties of reductive groups related to the existence of a (B,N)-pair. For an abstract group, having a (B,N)-pair is a very strong condition; many of the theorems we will give for reductive groups follow from this single property.

1.1 DEFINITION. *A (B,N)-**pair** in a group G is a system (B,N,S) which consists of two subgroups B and N such that $B \cap N$ is normal in N, and of a set S of involutions of the quotient group $W = N/(B \cap N)$ having the following properties:*
(i) The set $B \cup N$ generates G.
(ii) The set S generates W.
(iii) For any $s{\in}S$ and $w{\in}W$ we have $sBw \subset BwB \cup BswB$.
(iv) For any $s{\in}S$ we have $sBs \not\subset B$.

The group W is called the Weyl group of the (B,N)-pair. The elementary properties of (B,N)-pairs can be found in [Bbk, IV, §2]. It is shown there that W is a Coxeter group with set of generators S and that (iii) can be refined to

$$\text{(iii}'\text{)} \qquad BsBwB = \begin{cases} BswB & \text{if } l(sw) = l(w)+1, \\ BwB \cup BswB & \text{if } l(sw) = l(w)-1. \end{cases}$$

In what follows we will often write elements of W (instead of representatives of them in N) in expressions representing subsets of G when these expressions do not depend upon the chosen representative.

1.2 THEOREM. *In a connected reductive group $\mathbf{G}$ we get a (B,N)-pair by taking for B some Borel subgroup $\mathbf{B}$, for N the normalizer $N_{\mathbf{G}}(\mathbf{T})$ of some maximal torus $\mathbf{T}$ of $\mathbf{B}$, and for S the set of s_α, as in 0.44 (iii), where α runs over the basis of Φ corresponding to $\mathbf{B}$ (see 0.31 (v)).*

PROOF: First we prove that $\mathbf{B} \cap N_{\mathbf{G}}(\mathbf{T}) = \mathbf{T}$, which amounts to proving that $R_u(\mathbf{B}) \cap N_{\mathbf{G}}(\mathbf{T}) = \{1\}$. Suppose that $v \in R_u(\mathbf{B}) \cap N_{\mathbf{G}}(\mathbf{T})$; then for any $t \in \mathbf{T}$, we have $[v,t] \in R_u(\mathbf{B}) \cap \mathbf{T}$ so, since this last group is trivial, we get $v \in C_{\mathbf{G}}(\mathbf{T}) = \mathbf{T}$ (the latter equality by 0.32 (iii)) so $v \in R_u(\mathbf{B}) \cap \mathbf{T} = \{1\}$.

According to 0.31 $\mathbf{G}$ is generated by $\mathbf{T}$ and the corresponding subgroups $\mathbf{U}_\alpha$; according to 0.44 (i) the element s_α conjugates $\mathbf{U}_\alpha$ to $\mathbf{U}_{-\alpha}$. Thus $\mathbf{G}$

is generated by $\mathbf{T}$, $N_{\mathbf{G}}(\mathbf{T})$ and the $\mathbf{U}_\alpha$ for $\alpha > 0$, *i.e.*, by $\mathbf{B}$ and $N_{\mathbf{G}}(\mathbf{T})$, whence axiom (i) of (B, N)-pairs; proposition 0.28 gives (ii). We also get (iv) since if α is positive, $\mathbf{U}_{-\alpha} = s_\alpha \mathbf{U}_\alpha s_\alpha \subset s_\alpha \mathbf{B} s_\alpha$ is not in $\mathbf{B}$ (as $\mathbf{B}$ contains only the subgroups $\mathbf{U}_\alpha$ for positive α).

We show now (iii). We have $\mathbf{B} = \mathbf{T}(\prod_{\beta\in\Phi^+-\{\alpha\}} \mathbf{U}_\beta)\mathbf{U}_\alpha$ whence $s_\alpha \mathbf{B} w = \mathbf{T}(\prod_{\beta\in\Phi^+-\{\alpha\}} \mathbf{U}_\beta) s_\alpha \mathbf{U}_\alpha w \subset \mathbf{B} s_\alpha \mathbf{U}_\alpha w$ (since for $\alpha \in \Phi$ and $w \in W$ we have $w\mathbf{U}_\alpha w^{-1} = \mathbf{U}_{w(\alpha)}$ (see remarks after 0.44), and since $s_\alpha(\beta)$ is positive for $\beta \in \Phi^+ - \{\alpha\}$ (see 0.29)). If $w^{-1}(\alpha) > 0$ we get $s_\alpha \mathbf{U}_\alpha w = s_\alpha w \mathbf{U}_{w^{-1}(\alpha)} \subset sw\mathbf{B}$, whence the result in that case. Otherwise a computation in $\mathbf{SL}_2$ shows that $s_\alpha \mathbf{U}_\alpha s_\alpha \subset \mathbf{B} s_\alpha \mathbf{U}_\alpha \cup \mathbf{T}\mathbf{U}_\alpha$ whence $\mathbf{B} s_\alpha \mathbf{U}_\alpha w \subset \mathbf{B} s_\alpha \mathbf{U}_\alpha s_\alpha w \cup \mathbf{B}\mathbf{U}_\alpha s_\alpha w = \mathbf{B} w \mathbf{U}_{w^{-1}(-\alpha)} \cup \mathbf{B} s_\alpha w$ whence the result (since in that case $w^{-1}(-\alpha) > 0$ so $\mathbf{U}_{w^{-1}(-\alpha)} \subset \mathbf{B}$). ■

1.3 REMARK. When the base field k is the algebraic closure of a finite field $\mathbb{F}_q$ and $\mathbf{G}$, $\mathbf{B}$ and $\mathbf{T}$ are defined over $\mathbb{F}_q$, it can be shown (see [BT]) that the groups of points over $\mathbb{F}_q$ of $\mathbf{B}$ and of $N_{\mathbf{G}}(\mathbf{T})$ form a (B, N)-pair in the group of points over $\mathbb{F}_q$ of $\mathbf{G}$, where the corresponding Weyl group is the subgroup of elements of $W(\mathbf{T})$ fixed under the action of the Frobenius endomorphism (see 3.15 (vi) below).

1.4 PROPOSITION (BRUHAT DECOMPOSITION). *Let G be a group with a (B, N)-pair. Then $G = \coprod_W BwB$ where W is the Weyl group of the (B, N)-pair.*

PROOF: As B and N generate G, every element of G has a decomposition of the form $b_1 n_1 b_2 n_2 \ldots$ with $n_i \in N$ and $b_i \in B$. Using property (iii) in the definition of (B, N)-pairs, this product may be transformed to bwb' with $b, b' \in B$ and $w \in W$. It remains to show that the double cosets BwB and $Bw'B$ are disjoint if $w \neq w'$ (as they are cosets, they are either disjoint or equal). We will do this by induction on the length $l(w)$, assumed to be less than or equal to $l(w')$. Assume that $BwB = Bw'B$; if $l(w) = 0$, *i.e.*, $w = 1$, then $w' \in B$ so is equal to $1 \in W = N/(B \cap N)$; if $l(w) > 0$ then there exists $s \in S$ such that $l(sw) < l(w)$, and by 1.1 (iii′) we have $BsB \cup BswB = BsBwB = BsBw'B \subset Bsw'B \cup Bw'B$. The induction hypothesis applied to sw shows that $BswB$ is disjoint from $Bw'B$ unless $sw = w'$, which is impossible since $l(sw) < l(w) \leq l(w')$. So we must have $BswB = Bsw'B$, which implies by the induction hypothesis that $sw = sw'$, whence the result. ■

1.5 COROLLARY. *In a reductive group, the intersection of two Borel subgroups always contains a maximal torus, and the two Borel subgroups are conjugate by an element of the normalizer of that torus.*

PROOF: Let $\mathbf{B}$ and $\mathbf{B}'$ be two Borel subgroups of $\mathbf{G}$, and let $\mathbf{T}$ be a maximal torus of $\mathbf{B}$. Let us write $\mathbf{B}' = {}^g\mathbf{B}$ for some $g \in \mathbf{G}$; applying the Bruhat decomposition with respect to the (B,N)-pair $(\mathbf{B}, N_{\mathbf{G}}(\mathbf{T}))$ to g we get $g = bnb_1$, with $b, b_1 \in \mathbf{B}$ and $n \in N_{\mathbf{G}}(\mathbf{T})$. So $\mathbf{B}' = {}^{bn}\mathbf{B}$, and the torus ${}^{gb_1^{-1}}\mathbf{T} = {}^b\mathbf{T}$ is in $\mathbf{B}' \cap \mathbf{B}$. Furthermore $\mathbf{B}' = {}^{bnb^{-1}}\mathbf{B}$ and $bnb^{-1} \in N_{\mathbf{G}}({}^b\mathbf{T})$. ∎

1.6 PROPOSITION. *Let G be a group with a (B,N)-pair; the subgroups of G which contain B are the $P_J = BW_JB$ for some $J \subset S$, where W_J is the (Coxeter) subgroup of W generated by J.*

PROOF: See, *e.g.*, [Bbk, IV, 2.5 theorem 3] (the fact that BW_JB is a group follows from 1.1 (iii)). ∎

A subgroup of a Coxeter group of the form W_J for $J \subset S$ is called a **parabolic subgroup** of W.

Let $\mathbf{G}$ be a connected reductive group, $\mathbf{T}$ a maximal torus of $\mathbf{G}$, and $\mathbf{B} = \mathbf{U} \rtimes \mathbf{T}$ the corresponding semi-direct product decomposition of a Borel subgroup containing $\mathbf{T}$. We put $\mathbf{U}_w = \prod_{\{\alpha>0 \mid w(\alpha)<0\}} \mathbf{U}_\alpha$; it is a subgroup of $\mathbf{U}$ by 0.31 (iv). The next proposition is a refinement of the Bruhat decomposition for $\mathbf{G}$ which gives a unique product decomposition of double $\mathbf{B}$-cosets.

1.7 PROPOSITION. *Every element of $\mathbf{G}$ has a unique decomposition of the form unv with $u \in \mathbf{U}$, $n \in N_{\mathbf{G}}(\mathbf{T})$, and $v \in \mathbf{U}_w$, where w denotes the image of n in W.*

PROOF: By 1.4 any element has a decomposition of the form unu' with $u, u' \in \mathbf{U}$ and $n \in N_{\mathbf{G}}(\mathbf{T})$. We may decompose u' by decomposing $\mathbf{U}$ as $\mathbf{U} = (\mathbf{U} \cap {}^{w^{-1}}\mathbf{U})\mathbf{U}_w$ (which holds since $\mathbf{U} \cap {}^{w^{-1}}\mathbf{U} = \prod_{\{\alpha>0 \mid w(\alpha)>0\}} \mathbf{U}_\alpha$). If $u' = v'v$ in this decomposition, we get $unu' = u.{}^nv'nv$, and $u.{}^nv' \in \mathbf{U}$. To show the uniqueness of this decomposition we have to show that an element unv as in the statement is in $N_{\mathbf{G}}(\mathbf{T})$ only if $u = v = 1$. Suppose $unv = n' \in N_{\mathbf{G}}(\mathbf{T})$; then, by 1.4, n and n' have the same image w in W, so $n' = tn$ for some $t \in \mathbf{T}$. We get ${}^nv = u^{-1}t \in \mathbf{B} \cap {}^n\mathbf{U}_w$; but by definition of $\mathbf{U}_w$ this intersection is trivial, whence the result. ∎

Parabolic subgroups; Levi subgroups.

1.8 DEFINITION. *A closed subgroup of a connected algebraic group* **G** *which contains a Borel subgroup is called a* **parabolic subgroup**.

1.9 PROPOSITION. *If* **G** *is reductive and connected,* **T** *is a maximal torus of* **G** *and* **B** *a Borel subgroup containing* **T**, *then every parabolic subgroup is conjugate to a unique* $\mathbf{P}_J = \mathbf{B}W_J\mathbf{B}$ *(see 1.6).*

PROOF: The uniqueness is 0.12 (ii). The other assertions follow from 1.6 and the fact that all Borel subgroups of **G** are conjugate. ■

Thus **G**-conjugacy classes of parabolic subgroups are parametrized by the subsets of the basis of Φ.

In the remainder of this chapter **G** will denote a connected reductive algebraic group. Let **T** be a maximal torus of **G**, and $\mathbf{B} = \mathbf{U} \rtimes \mathbf{T}$ be a Borel subgroup which contains **T**; let Φ be the set of roots of **G** relative to **T** and Π the basis of Φ corresponding to **B**. For $I \subset \Pi$ let Φ_I be the set of roots which are in the subspace of $X(\mathbf{T}) \otimes \mathbb{R}$ generated by I. It is easy to check that Φ_I is a root system in that subspace, and that I is a basis of Φ_I.

1.10 LEMMA. *For* $\alpha, \beta \in \Phi^+$ *let* m *and* n *be positive integers such that* $m\alpha + n\beta$ *is a root; then* $m\alpha + n\beta \in \Phi_I$ *if and only if* $\alpha, \beta \in \Phi_I$.

PROOF: The lemma immediately follows from the remark that every root of Π which occurs in the decomposition of α or β as a combination with positive coefficients of roots of Π also occurs in the decomposition of $m\alpha + n\beta$. ■

As above, we will denote by $\mathbf{P}_I$ the parabolic subgroup $\mathbf{B}W_I\mathbf{B}$.

1.11 PROPOSITION. *The unipotent radical of* $\mathbf{P}_I$ *is* $\mathbf{V}_I = \prod_{\alpha \in \Phi^+ - \Phi_I} \mathbf{U}_\alpha$.

PROOF: It is clear by 0.31 (iv) and the lemma that $\mathbf{V}_I$ is a group, normalized by all U_α with $\alpha \in \Phi^+$.. It is also normalized by **T**; so to show that it is normalized by $\mathbf{P}_I$ it is enough to show that it is normalized by W_I. Since any element of W_I is a product of s_α with $\alpha \in I$ and ${}^{s_\alpha}\mathbf{U}_\beta = \mathbf{U}_{s_\alpha(\beta)}$, we just have to show that if $\beta \in \Phi^+ - \Phi_I$ and $\alpha \in I$ then $s_\alpha(\beta) \in \Phi^+ - \Phi_I$. Since $l(s_\alpha) = 1$ the only positive root sent to a negative root by s_α is α, so $s_\alpha(\beta)$ is positive, and it cannot be in Φ_I since $s_\alpha(\Phi_I) = \Phi_I$.

The unipotent radical of an algebraic group is in all Borel subgroups (see 0.32 (i)) so $R_u(\mathbf{P}_I) \subset \mathbf{U}$, and, as it is normalized by **T**, it is a product of the $\mathbf{U}_\alpha$ it contains. Since $\mathbf{V}_I$ is normal in $\mathbf{P}_I$ it is in $R_u(\mathbf{P}_I)$, so $R_u(\mathbf{P}_I)$ is of the form $\prod_{\alpha \in \Psi} \mathbf{U}_\alpha$ where $\Phi^+ - \Phi_I \subset \Psi \subset \Phi^+$. To finish the proof, let us show that if $\alpha \in \Phi_I$ then $\mathbf{U}_\alpha$ cannot be in $R_u(\mathbf{P}_I)$: otherwise, since $s_\alpha \in W_I$, the group $\mathbf{U}_{-\alpha} = {}^{s_\alpha}\mathbf{U}_\alpha$ would also be in $R_u(\mathbf{P}_I)$, but this is impossible since $<\mathbf{U}_\alpha, \mathbf{U}_{-\alpha}>$ contains semi-simple elements (as a simple computation in $\mathbf{SL}_2$ shows; see 0.44). ■

We will now show that the reductive quotient of $\mathbf{P}_I$ can be lifted back. To do that, and more generally to study reductive subgroups of $\mathbf{G}$ of maximal rank (the **rank** of an algebraic group is the dimension of its maximal tori), we need the following definition.

1.12 DEFINITION.

(i) *A subset $\Psi \subset \Phi$ is called* **quasi-closed** *if the group $\mathbf{G}^*_\Psi = <\mathbf{U}_\alpha | \alpha \in \Psi>$ does not contain any $\mathbf{U}_\gamma$ where γ is a root not in Ψ.*

(ii) *A subset Ψ of a root system is said to be* **closed** *if whenever $\alpha, \beta \in \Psi$, and n and m are positive integers such that $n\alpha + m\beta$ is a root, then $n\alpha + m\beta \in \Psi$.*

1.13 PROPOSITION. *A closed subset of Φ is quasi-closed.*

PROOF: See [BT, 3.4]. ■

Conversely any quasi-closed subset of Φ is closed apart from some exceptions in characteristics 2 and 3 (see [BT, 3.8]).

A quasi-closed and symmetric subset is a root system in the subspace it generates. This is for instance a corollary of the following proposition.

1.14 PROPOSITION. *The closed and connected subgroups of $\mathbf{G}$ which contain $\mathbf{T}$ are the $\mathbf{G}_\Psi = <\mathbf{T}, \mathbf{U}_\alpha \mid \alpha \in \Psi>$, where Ψ is a quasi-closed subset of Φ. Furthermore, if $\mathbf{U}_\alpha \subset \mathbf{G}_\Psi$ then $\alpha \in \Psi$. The group $\mathbf{G}_\Psi$ is reductive if and only if Ψ is symmetric.*

PROOF: By 0.34 such a subgroup is generated by $\mathbf{T}$ and the $\mathbf{U}_\alpha$ it contains. The subset $\Psi \subset \Phi$ of those $\mathbf{U}_\alpha$ is quasi-closed by definition. If $\mathbf{G}_\Psi$ is reductive, as the root system of a reductive group, Ψ is symmetric. Conversely, if Ψ is a quasi-closed subset of the root system Φ of $\mathbf{G}$, the group $\mathbf{G}_\Psi$ is closed and connected by 0.2; let us show that it is reductive if Ψ is symmetric.

Let $\mathbf{V}$ be its unipotent radical; since it is normal in $\mathbf{G}_\Psi$ it is generated by the $\mathbf{U}_\alpha$ it contains (see 0.33). As the quotient $\mathbf{G}_\Psi/\mathbf{G}^*_\Psi$ is a torus (a quotient of $\mathbf{T}$), every $\mathbf{U}_\alpha \subset \mathbf{V}$ is in $\mathbf{G}^*_\Psi$, and, as Ψ is quasi-closed, the root α is in Ψ. We finish the proof as in 1.11: if $\mathbf{U}_\alpha \subset \mathbf{V}$ and $\alpha \in \Psi$, as Ψ is symmetric, we have $\mathbf{U}_{-\alpha} \in \mathbf{G}_\Psi$, whence, because $\mathbf{V}$ is normal, we get $[\mathbf{U}_\alpha, \mathbf{U}_{-\alpha}] \subset \mathbf{V}$ which is impossible since this group contains semi-simple elements. The last part of the proposition comes from the definition of a quasi-closed set and the remark that any $\mathbf{U}_\alpha \subset \mathbf{G}_\Psi$ is in $\mathbf{G}^*_\Psi$. ■

1.15 PROPOSITION. *The group $\mathbf{L}_I = \mathbf{G}_{\Phi_I}$ is reductive, the set Φ_I is the set of roots of $\mathbf{L}_I$ relative to $\mathbf{T}$, and there is a semi-direct product decomposition $\mathbf{P}_I = R_u(\mathbf{P}_I) \rtimes \mathbf{L}_I$.*

PROOF: Let us write $\mathbf{U} = \mathbf{U}_I\mathbf{V}_I$ where $\mathbf{U}_I = \prod_{\alpha\in\Phi_I^+} \mathbf{U}_\alpha$ (see 0.33). Since $\mathbf{V}_I$ is normal in $\mathbf{P}_I$ and $\mathbf{P}_I = \mathbf{B}W_I\mathbf{B}$ we have $\mathbf{P}_I = \mathbf{U}_I\mathbf{T}W_I\mathbf{U}_I\mathbf{V}_I \subset \mathbf{L}_I\mathbf{V}_I$ ($\mathbf{T}W_I \subset \mathbf{L}_I$ since if $\alpha \in \Phi_I$ then s_α has a representative in $<\mathbf{U}_\alpha, \mathbf{U}_{-\alpha}> \subset \mathbf{L}_I$). Furthermore, as by definition the set Φ_I is closed, from 1.14 $\mathbf{L}_I$ is reductive and contains only those $\mathbf{U}_\alpha$ for which $\alpha \in \Phi_I$, so $\mathbf{L}_I \cap \mathbf{V}_I = 1$ and we have $\mathbf{P}_I = \mathbf{V}_I \rtimes \mathbf{L}_I$. ■

1.16 DEFINITION. *An algebraic group $\mathbf{P}$ is said to have a* **Levi decomposition** *if there is a closed subgroup $\mathbf{L} \subset \mathbf{P}$ such that $\mathbf{P} = R_u(\mathbf{P}) \rtimes \mathbf{L}$. The group $\mathbf{L}$ is called a* **Levi subgroup** *of $\mathbf{P}$ (or a* **Levi complement***).*

We have just seen that a parabolic subgroup has a Levi decomposition.

1.17 PROPOSITION. *Let $\mathbf{P}$ be a parabolic subgroup of $\mathbf{G}$, and $\mathbf{T}$ be a maximal torus of $\mathbf{P}$. There exists a unique Levi subgroup of $\mathbf{P}$ containing $\mathbf{T}$.*

PROOF: Let $\mathbf{B}$ be a Borel subgroup of $\mathbf{P}$ which contains $\mathbf{T}$; by 1.6 $\mathbf{P}$ is a $\mathbf{P}_I$, and the above proposition shows the existence of a Levi subgroup (the subgroup $\mathbf{L}_I$) containing $\mathbf{T}$.

Conversely, a Levi subgroup $\mathbf{L} \subset \mathbf{P}_I$ containing $\mathbf{T}$ is generated by $\mathbf{T}$ and the $\mathbf{U}_\alpha$ it contains. If $\alpha \in \Phi^+ - \Phi_I$, then $\mathbf{U}_\alpha \subset R_u(\mathbf{P}_I)$ so is not in $\mathbf{L}$. As the only α such that $\mathbf{U}_\alpha \subset \mathbf{P}_I$ are the roots in $\Phi_I \cup \Phi^+$, we get that $\mathbf{L}$ is in $\mathbf{L}_I$, so must be equal to it. ■

1.18 COROLLARY. *Two Levi subgroups of a parabolic subgroup $\mathbf{P}$ are conjugate by a unique element of $R_u(\mathbf{P})$.*

PROOF: Let $\mathbf{L}$ and $\mathbf{L}'$ be two Levi subgroups of $\mathbf{P}$. Let $\mathbf{T}$ (resp. $\mathbf{T}'$) be a maximal torus of $\mathbf{L}$ (resp. $\mathbf{L}'$). As maximal tori of $\mathbf{P}$, the groups $\mathbf{T}$ and $\mathbf{T}'$ are conjugate by an element of $\mathbf{P}$ which may be written as vl with $v \in R_u(\mathbf{P})$ and $l \in \mathbf{L}$. By 1.17 vl also conjugates $\mathbf{L}$ to $\mathbf{L}'$ so ${}^{vl}\mathbf{L} = {}^{v}\mathbf{L} = \mathbf{L}'$. It remains to show the uniqueness of v, which amounts to showing that $N_{\mathbf{G}}(\mathbf{L}) \cap R_u(\mathbf{P}) = 1$. Suppose that $v \in R_u(\mathbf{P}) \cap N_{\mathbf{G}}(\mathbf{L})$; then, for any $l \in \mathbf{L}$, we have $[v, l] \in R_u(\mathbf{P}) \cap \mathbf{L}$ so, since this last group is trivial by 1.15, we get $v \in C_{\mathbf{G}}(\mathbf{L})$. But $C_{\mathbf{G}}(\mathbf{L}) = Z(\mathbf{L})$ (since, *e.g.*, by 0.32 (iii) we have

$C_{\mathbf{G}}(\mathbf{L}) \subset \mathbf{L}$) so v must be equal to 1 (compare with the beginning of the proof of 1.2). ■

1.19 PROPOSITION. *Let* $\mathbf{L}$ *be a Levi subgroup of a parabolic subgroup* $\mathbf{P}$. *Then* $R(\mathbf{P}) = R_u(\mathbf{P}) \rtimes R(\mathbf{L})$.

PROOF: The quotient $\mathbf{P}/R(\mathbf{L})R_u(\mathbf{P})$ is isomorphic to $\mathbf{L}/R(\mathbf{L})$, so is semi-simple. So $R(\mathbf{P}) \subset R(\mathbf{L})R_u(\mathbf{P})$. But $R(\mathbf{L})R_u(\mathbf{P})$ is connected, solvable and normal in $\mathbf{P}$ as the inverse image of a normal subgroup of the quotient $\mathbf{P}/R_u(\mathbf{P}) \simeq \mathbf{L}$, whence the reverse inclusion. ■

We will now characterize parabolic subgroups in terms of roots.

1.20 PROPOSITION.

(i) *A closed subgroup* $\mathbf{P}$ *of* $\mathbf{G}$ *which contains* $\mathbf{T}$ *is parabolic if and only if for any root* $\alpha \in \Phi$, *either* $\mathbf{U}_\alpha \subset \mathbf{P}$ *or* $\mathbf{U}_{-\alpha} \subset \mathbf{P}$.

(ii) *A subset* $\Psi \subset \Phi$ *is the set of* α *such that* $\mathbf{U}_\alpha$ *is in a given parabolic subgroup if and only if there exists a vector* x *of* $X(\mathbf{T}) \otimes \mathbb{R}$ *such that* $\Psi = \{\alpha \in \Phi \mid \langle \alpha, x \rangle \geq 0\}$. *The set of roots relative to* $\mathbf{T}$ *of the Levi subgroup containing* $\mathbf{T}$ *of that parabolic subgroup is* $\{\alpha \in \Phi \mid \langle \alpha, x \rangle = 0\}$.

PROOF: We have seen that (i) holds for a parabolic subgroup (see, *e.g.*, 1.11 and 1.15). Conversely, let $\mathbf{P}$ be such that for any α either $\mathbf{U}_\alpha$ or $\mathbf{U}_{-\alpha}$ is in $\mathbf{P}$. To show that $\mathbf{P}$ is a parabolic subgroup it is enough to show that it contains a Borel subgroup of $\mathbf{G}$. Let $\mathbf{B}'$ be a Borel subgroup of $\mathbf{P}$ and $\mathbf{B}$ be a Borel subgroup of $\mathbf{G}$ containing $\mathbf{B}'$; we will show that $\mathbf{B} = \mathbf{B}'$. Let α be a positive root for the order on Φ defined by $\mathbf{B}$. We show that $\mathbf{U}_\alpha \subset \mathbf{B}'$. Suppose first that $\mathbf{U}_\alpha \subset \mathbf{P}$; then $<\mathbf{U}_\alpha, \mathbf{B}'>$ is a subgroup of $\mathbf{B} \cap \mathbf{P}$ so is equal to $\mathbf{B}'$, whence the result. Otherwise $\mathbf{U}_{-\alpha} \subset \mathbf{P}$. We proceed by contradiction; if $\mathbf{U}_\alpha \not\subset \mathbf{B}'$ then $\mathbf{U}_\alpha \not\subset R_u(\mathbf{P})$ so $\mathbf{U}_\alpha$ maps isomorphically to a root subgroup, relative to the image of $\mathbf{T}$, of the reductive group $\mathbf{P}/R_u(\mathbf{P})$. Then $-\alpha$ is also a root of this group. Let n be a representative in $\mathbf{P}$ of the reflection $s_{-\alpha}$ in the Weyl group of $\mathbf{P}/R_u(\mathbf{P})$; the element n conjugates $\mathbf{T}$ to another torus of the group $\mathbf{T}.R_u(\mathbf{P})$, so, as all maximal tori in this group are conjugate, n can be changed to another representative n' which normalizes $\mathbf{T}$. We note then that on ${}^{n'}\mathbf{U}_{-\alpha}$ the group $\mathbf{T}$ acts by $s_{-\alpha}(-\alpha) = \alpha$ which contradicts $\mathbf{U}_\alpha \not\subset \mathbf{P}$.

We show now (ii). A set of the form $\Psi = \{\alpha \in \Phi \mid \langle \alpha, x \rangle \geq 0\}$ is clearly closed, so is the set of $\mathbf{U}_\alpha$ in the group $\mathbf{G}_\Psi$; and this group is parabolic as Ψ obviously satisfies the condition in (i). By conjugating $\mathbf{G}_\Psi$ to some $\mathbf{P}_I$, we see that its Levi subgroup containing $\mathbf{T}$ also contains the $\mathbf{U}_\alpha$ for which both α and $-\alpha$ are in Ψ, which is the set $\{\alpha \in \Phi \mid \langle \alpha, x \rangle = 0\}$.

Conversely, we may assume, by conjugating if necessary by $W(\mathbf{T})$, that the parabolic subgroup we consider is $\mathbf{P}_I$, so $\Psi = \Phi^+ \cup \Phi_I$. Take any x such that $\langle x, \alpha \rangle = 0$ if $\alpha \in \Phi_I$ and $\langle x, \alpha \rangle > 0$ if $\alpha \in \Psi - \Phi_I$. Such an x exists: the projection of Φ^+ on $\Phi_I^\perp$ lies in a half-space, and we may take x in this half-space, orthogonal to the hyperplane which delimits it. It is clear that by construction x has the required properties. ■

We now give an important property of Levi subgroups.

1.21 PROPOSITION. *Let* $\mathbf{L}$ *be a Levi subgroup of a parabolic subgroup of* $\mathbf{G}$*; then* $\mathbf{L} = C_{\mathbf{G}}(Z(\mathbf{L})^\circ)$.

PROOF: We may assume we are in the above situation with $\mathbf{L} = \mathbf{L}_I$. Then by 0.35 the group $Z(\mathbf{L})$ is the intersection of the kernels of the roots in Φ_I. The group $C_{\mathbf{G}}(Z(\mathbf{L})^\circ)$ is connected as it is the centralizer of a torus (see 0.6 and 0.11 (iii)) and it is generated by $\mathbf{T}$ and the $\mathbf{U}_\alpha$ it contains (see 0.34; it is normalized by $\mathbf{T}$ because it contains $\mathbf{T}$); the $\mathbf{U}_\alpha$ it contains are such that α is trivial on $(\bigcap_{\alpha\in\Phi_I} \ker\alpha)^\circ$. Since the identity component above is of finite index, this implies that some multiple $n\alpha$ of α is trivial on $\bigcap_{\alpha\in\Phi_I} \ker\alpha$. With the notation of 0.23, this can be rewritten as $n\alpha \in (<\Phi_I>^\perp)^\perp$. But $(<\Phi_I>^\perp)^\perp/<\Phi_I>$ is a torsion group (see 0.24). This implies that $\alpha \in <\Phi_I> \otimes \mathbb{Q}$, which in turn yields $\alpha \in \Phi_I$ by the definition of Φ_I. This proves that $C_{\mathbf{G}}(Z(\mathbf{L})^\circ) \subset \mathbf{L}$. The reverse inclusion is obvious. ■

The next proposition is a kind of converse.

1.22 PROPOSITION. *For any torus* $\mathbf{S}$*, the group* $C_{\mathbf{G}}(\mathbf{S})$ *is a Levi subgroup of some parabolic subgroup of* $\mathbf{G}$.

PROOF: Let $\mathbf{T}$ be a maximal torus containing $\mathbf{S}$. As the group $C_{\mathbf{G}}(\mathbf{S})$ is connected by 0.11 (iii) and contains $\mathbf{T}$, by 0.34 we have $C_{\mathbf{G}}(\mathbf{S}) = <\mathbf{T}, \mathbf{U}_\alpha \mid \mathbf{U}_\alpha \subset C_{\mathbf{G}}(\mathbf{S})>$. As $\mathbf{S}$ acts by α on $\mathbf{U}_\alpha$ (see 0.31 (i)), we have

$$\mathbf{U}_\alpha \subset C_{\mathbf{G}}(\mathbf{S}) \Leftrightarrow \alpha|_{\mathbf{S}} = 0,$$

where 0 is the trivial element of $X(\mathbf{S})$. Let us choose a total order on $X(\mathbf{S})$ (*i.e.*, a structure of ordered $\mathbb{Z}$-module). As $X(\mathbf{S})$ is a quotient of $X(\mathbf{T})$ (see 0.21) there exists a total order on $X(\mathbf{T})$ compatible with the chosen order on $X(\mathbf{S})$, *i.e.*, such that for $x \in X(\mathbf{T})$ we have $x \geq 0 \Rightarrow x|_{\mathbf{S}} \geq 0$. This implies that the set $\Psi = \{\alpha \in \Phi \mid \alpha > 0 \text{ or } \alpha|_{\mathbf{S}} = 0\}$ is also equal to $\{\alpha \in \Phi \mid \alpha|_{\mathbf{S}} \geq 0\}$. This last definition implies that Ψ is closed, so (see 1.13 and 1.14) Ψ is also the set of α such that $\mathbf{U}_\alpha \subset \mathbf{G}_\Psi$. It follows then from 1.20 that $\mathbf{G}_\Psi$ is a parabolic subgroup, of which $C_{\mathbf{G}}(\mathbf{S})$ is a Levi complement. ■

References

For further elementary properties of (B, N)-pairs, see chapter IV of [Bbk]. For additional properties of closed and quasi-closed subsets of a root system, see [BT].

2. INTERSECTIONS OF PARABOLIC SUBGROUPS, REDUCTIVE SUBGROUPS OF MAXIMAL RANK, CENTRALIZERS OF SEMI-SIMPLE ELEMENTS

We first study the intersection of two parabolic subgroups. First note that by 1.5 the intersection of two parabolic subgroups always contains some maximal torus of $\mathbf{G}$.

2.1 PROPOSITION. *Let $\mathbf{P}$ and $\mathbf{Q}$ be two parabolic subgroups of $\mathbf{G}$ whose unipotent radicals are respectively $\mathbf{U}$ and $\mathbf{V}$, and let $\mathbf{L}$ and $\mathbf{M}$ be Levi subgroups of $\mathbf{P}$ and $\mathbf{Q}$ respectively containing the same maximal torus of $\mathbf{G}$; then:*

(i) *The group $(\mathbf{P}\cap\mathbf{Q}).\mathbf{U}$ is a parabolic subgroup of $\mathbf{G}$ included in $\mathbf{P}$; it has the same intersection as $\mathbf{Q}$ with $\mathbf{L}$, and it has $\mathbf{L}\cap\mathbf{M}$ as a Levi subgroup.*

(ii) *The group $\mathbf{P}\cap\mathbf{Q}$ is connected.*

(iii) *The group $\mathbf{P}\cap\mathbf{Q}$ has the following product decomposition:*

$$\mathbf{P}\cap\mathbf{Q} = (\mathbf{L}\cap\mathbf{M}).(\mathbf{L}\cap\mathbf{V}).(\mathbf{M}\cap\mathbf{U}).(\mathbf{U}\cap\mathbf{V}),$$

which is a product of varieties; in particular the decomposition of an element of $\mathbf{P}\cap\mathbf{Q}$ as a product of four terms is unique.

PROOF: Let $\mathbf{T}$ be a maximal torus of $\mathbf{L}\cap\mathbf{M}$, and α be a root of $\mathbf{G}$ relative to $\mathbf{T}$. By 1.20 (i) to prove the first part of (i) it is enough to show that either $\mathbf{U}_\alpha$ or $\mathbf{U}_{-\alpha}$ is in $(\mathbf{P}\cap\mathbf{Q}).\mathbf{U}$. If neither $\mathbf{U}_\alpha$ nor $\mathbf{U}_{-\alpha}$ is in $\mathbf{U}$, then they are both in $\mathbf{L}$. As one of them is in $\mathbf{Q}$, it is in $\mathbf{L}\cap\mathbf{Q}$, so in $(\mathbf{P}\cap\mathbf{Q}).\mathbf{U}$.

We now prove (ii). By 0.12 (i) $(\mathbf{P}\cap\mathbf{Q}).\mathbf{U}$ is connected so $((\mathbf{P}\cap\mathbf{Q}).\mathbf{U})/\mathbf{U}$ which equals $(\mathbf{P}\cap\mathbf{Q})/(\mathbf{U}\cap\mathbf{Q})$ is also connected. As $\mathbf{U}\cap\mathbf{Q}$ is connected by 0.33, so is in $(\mathbf{P}\cap\mathbf{Q})^\circ$, this implies that $\mathbf{P}\cap\mathbf{Q}$ is connected.

Let us prove (iii). By (ii) and 0.34, we have

$$\mathbf{P}\cap\mathbf{Q} = <\mathbf{T},\mathbf{U}_\alpha \mid \mathbf{U}_\alpha\subset\mathbf{P}\cap\mathbf{Q}>.$$

Let α be a root from the set above. From what we know about parabolic subgroups, we are in one of the following cases:

- $\mathbf{U}_{-\alpha}$ is neither in $\mathbf{P}$ nor in $\mathbf{Q}$. In that case $\mathbf{U}_\alpha\subset\mathbf{U}\cap\mathbf{V}$.
- $\mathbf{U}_{-\alpha}$ is in $\mathbf{P}$ but not in $\mathbf{Q}$ (resp. in $\mathbf{Q}$ and not in $\mathbf{P}$). In that case $\mathbf{U}_\alpha\subset\mathbf{L}\cap\mathbf{V}$ (resp. $\mathbf{U}_\alpha\subset\mathbf{M}\cap\mathbf{U}$).

- $\mathbf{U}_{-\alpha}$ is in $\mathbf{P} \cap \mathbf{Q}$. In that case $\mathbf{U}_\alpha \subset \mathbf{L} \cap \mathbf{M}$.

We thus get

$$\mathbf{P} \cap \mathbf{Q} = \langle \mathbf{U} \cap \mathbf{V}, \mathbf{L} \cap \mathbf{V}, \mathbf{M} \cap \mathbf{U}, \mathbf{L} \cap \mathbf{M} \rangle.$$

But $\mathbf{U} \cap \mathbf{V}$ is normal in $\mathbf{P} \cap \mathbf{Q}$, and similarly $\mathbf{L} \cap \mathbf{M}$ normalizes $\mathbf{L} \cap \mathbf{V}$ and $\mathbf{M} \cap \mathbf{U}$. Thus

$$\mathbf{P} \cap \mathbf{Q} = (\mathbf{L} \cap \mathbf{M}).\langle \mathbf{L} \cap \mathbf{V}, \mathbf{M} \cap \mathbf{U} \rangle.(\mathbf{U} \cap \mathbf{V}).$$

Furthermore, the commutator of an element of $\mathbf{L} \cap \mathbf{V}$ with an element of $\mathbf{M} \cap \mathbf{U}$ is in $\mathbf{U} \cap \mathbf{V}$. So we get the above-stated decomposition

$$\mathbf{P} \cap \mathbf{Q} = (\mathbf{L} \cap \mathbf{M}).(\mathbf{L} \cap \mathbf{V}).(\mathbf{M} \cap \mathbf{U}).(\mathbf{U} \cap \mathbf{V}).$$

Suppose now that $x = l_{\mathbf{M}} l_{\mathbf{V}} m_{\mathbf{U}} u_{\mathbf{V}} \in \mathbf{P} \cap \mathbf{Q}$, where $l_{\mathbf{M}} \in \mathbf{L} \cap \mathbf{M}$, $l_{\mathbf{V}} \in \mathbf{L} \cap \mathbf{V}$, *etc.* Then $l_{\mathbf{M}} l_{\mathbf{V}}$ is the image of x by the morphism $\mathbf{P} \to \mathbf{L}$ and $l_{\mathbf{M}}$ (resp. $m_{\mathbf{U}}$) is the image of $l_{\mathbf{M}} l_{\mathbf{V}}$ (resp. $m_{\mathbf{U}} u_{\mathbf{V}}$) by the morphism $\mathbf{Q} \to \mathbf{M}$. So the product map

$$(\mathbf{L} \cap \mathbf{M}) \times (\mathbf{L} \cap \mathbf{V}) \times (\mathbf{M} \cap \mathbf{U}) \times (\mathbf{U} \cap \mathbf{V}) \to \mathbf{P} \cap \mathbf{Q}$$

is an isomorphism of varieties. Note that since $\mathbf{P} \cap \mathbf{Q}$ is connected each of the 4 terms in the above product is connected.

It remains to prove the second part of (i). From the above decomposition we get $(\mathbf{P} \cap \mathbf{Q}).\mathbf{U} = (\mathbf{L} \cap \mathbf{M}).(\mathbf{L} \cap \mathbf{V}).\mathbf{U}$, whence $((\mathbf{P} \cap \mathbf{Q}).\mathbf{U}) \cap \mathbf{L} = (\mathbf{L} \cap \mathbf{M}).(\mathbf{L} \cap \mathbf{V}) = \mathbf{Q} \cap \mathbf{L}$. Now, as $\mathbf{L} \cap \mathbf{M}$ normalizes $\mathbf{L} \cap \mathbf{V}$ and $\mathbf{U}$, we get that $(\mathbf{L} \cap \mathbf{V}).\mathbf{U}$ is a normal subgroup of $(\mathbf{P} \cap \mathbf{Q}).\mathbf{U}$. But $\mathbf{L} \cap \mathbf{V}$ is connected, unipotent and normalizes $\mathbf{U}$, thus the product $(\mathbf{L} \cap \mathbf{V}).\mathbf{U}$ is unipotent and connected, so in $R_u((\mathbf{P} \cap \mathbf{Q}).\mathbf{U})$. The group $\mathbf{L} \cap \mathbf{M}$ is connected, normalized by $\mathbf{T}$, and so generated by $\mathbf{T}$ and the $\mathbf{U}_\alpha$ it contains. The set of such α is closed and symmetric as the intersection of two closed and symmetric sets, so by 1.14 this group is reductive, thus is a Levi complement of $(\mathbf{P} \cap \mathbf{Q}).\mathbf{U}$. ■

2.2 PROPOSITION. *Let* $\mathbf{H}$ *be a closed reductive subgroup of* $\mathbf{G}$ *of maximal rank. Then:*

(i) The Borel subgroups of $\mathbf{H}$ *are the* $\mathbf{B} \cap \mathbf{H}$ *where* $\mathbf{B}$ *is a Borel subgroup of* $\mathbf{G}$ *containing a maximal torus of* $\mathbf{H}$.

(ii) The parabolic subgroups of $\mathbf{H}$ *are the* $\mathbf{P} \cap \mathbf{H}$, *where* $\mathbf{P}$ *is a parabolic subgroup of* $\mathbf{G}$ *containing a maximal torus of* $\mathbf{H}$.

(iii) *If* $\mathbf{P}$ *is a parabolic subgroup of* $\mathbf{G}$ *containing a maximal torus of* $\mathbf{H}$, *the Levi subgroups of* $\mathbf{P}\cap\mathbf{H}$ *are the* $\mathbf{L}\cap\mathbf{H}$ *where* $\mathbf{L}$ *is a Levi subgroup of* $\mathbf{P}$ *containing a maximal torus of* $\mathbf{H}$.

PROOF: Let $\mathbf{T}$ be a maximal torus of $\mathbf{H}$; by assumption, it is also a maximal torus of $\mathbf{G}$. Let $\mathbf{B}$ be a Borel subgroup of $\mathbf{G}$ containing $\mathbf{T}$, and let $\mathbf{B} = \mathbf{U}.\mathbf{T}$ be the corresponding semi-direct product decomposition. The Borel subgroup $\mathbf{B}$ defines an order on the root system Φ (resp. $\Phi_{\mathbf{H}}$) of $\mathbf{G}$ (resp. $\mathbf{H}$) with respect to $\mathbf{T}$. The group $\mathbf{U}\cap\mathbf{H}$ is normalized by $\mathbf{T}$, so is connected and equal to the product of the $\mathbf{U}_\alpha$ it contains. It certainly contains those $\mathbf{U}_\alpha$ such that α is positive and in $\Phi_{\mathbf{H}}$, so $(\mathbf{U}\cap\mathbf{H}).\mathbf{T} = \mathbf{B}\cap\mathbf{H}$ is a Borel subgroup of $\mathbf{H}$. This gives (i) since all Borel subgroups of $\mathbf{H}$ are conjugate under $\mathbf{H}$.

Let us prove (ii). If $\mathbf{P}$ is a parabolic subgroup of $\mathbf{G}$ containing $\mathbf{T}$, it contains a Borel subgroup $\mathbf{B}$ containing $\mathbf{T}$, so its intersection with $\mathbf{H}$ contains the Borel subgroup $\mathbf{B}\cap\mathbf{H}$ of $\mathbf{H}$ and thus is a parabolic subgroup. Conversely, let $\mathbf{Q}$ be a parabolic subgroup of $\mathbf{H}$ containing $\mathbf{T}$, and let x be a vector of $X(\mathbf{T})\otimes\mathbb{R}$ defining $\mathbf{Q}$ as in 1.20 (ii). Then x defines a parabolic subgroup $\mathbf{P}$ of $\mathbf{G}$. It remains to show that $\mathbf{P}\cap\mathbf{H} = \mathbf{Q}$. The group $\mathbf{P}\cap\mathbf{H}$ is a parabolic subgroup of $\mathbf{H}$ by the first part. It is generated by $\mathbf{T}$ and the $\mathbf{U}_\alpha$ it contains. But $\mathbf{U}_\alpha \subset \mathbf{P}\cap\mathbf{H}$ if and only if $\alpha\in\Phi_{\mathbf{H}}$ and $\langle \alpha, x\rangle \geq 0$, *i.e.*, if and only if $\mathbf{U}_\alpha\subset\mathbf{Q}$ by definition of x.

Similarly, the Levi subgroup of $\mathbf{Q}$ containing $\mathbf{T}$ is the intersection of the Levi subgroup of $\mathbf{P}$ containing $\mathbf{T}$ with $\mathbf{H}$, as it is generated by $\mathbf{T}$ and the $\mathbf{U}_\alpha$ with $\alpha\in\Phi_{\mathbf{H}}$ orthogonal to x, whence (iii). ■

In the final part of this chapter we study centralizers of semi-simple elements.

2.3 PROPOSITION. *Let* $s\in\mathbf{G}$ *be a semi-simple element, and let* $\mathbf{T}$ *be a maximal torus containing* s; *then:*

(i) *The group* $C_{\mathbf{G}}(s)$ *is generated by the* $\mathbf{U}_\alpha$ *such that* $\alpha(s) = 1$, *and the elements* $n\in N_{\mathbf{G}}(\mathbf{T})$ *such that* ${}^n s = s$.

(ii) *The identity component* $C_{\mathbf{G}}(s)^\circ$ *is generated by* $\mathbf{T}$ *and the* $\mathbf{U}_\alpha$ *such that* $\alpha(s) = 1$. *It is a connected reductive subgroup of* $\mathbf{G}$ *of maximal rank.*

PROOF: Let $\mathbf{B} = \mathbf{T}\mathbf{U}$ be the Levi decomposition of a Borel subgroup of $\mathbf{G}$ which contains $\mathbf{T}$ and let $g\in C_{\mathbf{G}}(s)$; by the Bruhat decomposition (see 1.7) the element g has a unique decomposition of the form $g = unv$ with $n\in N_{\mathbf{G}}(\mathbf{T})$, $u\in\mathbf{U}$ and $v\in\mathbf{U}_w$ where w is the image of n in $W(\mathbf{T})$.

As $s \in \mathbf{T}$ and $\mathbf{T}$ normalizes $\mathbf{U}$ and $\mathbf{U}_w$, this decomposition is invariant under conjugation by s, so as ${}^s g = g$ we must have that each of u, n and v also centralizes s. Writing again unique decompositions of the form $u = \prod_{\alpha>0} u_\alpha$ and $v = \prod_{\alpha>0, {}^w\alpha<0} v_\alpha$ (where $u_\alpha, v_\alpha \in \mathbf{U}_\alpha$), we find once more that s centralizes u and v only if it centralizes all u_α and v_α. As s acts on $\mathbf{U}_\alpha$ by $\alpha(s)$ (see 0.31), we must have $\alpha(s) = 1$ for any α such that $u_\alpha \neq 1$ or $v_\alpha \neq 1$, whence (i).

The group $<\mathbf{T}, \mathbf{U}_\alpha \mid \alpha(s) = 1>$ is a connected and reductive subgroup of $\mathbf{G}$ (of maximal rank) by 1.14, since the set of α such that $\alpha(s) = 1$ is clearly closed and symmetric. It is normal in $C_{\mathbf{G}}(s)$ since if $n \in N_{\mathbf{G}}(\mathbf{T})$ centralizes s and has image w in $W(\mathbf{T})$, then ${}^n\mathbf{U}_\alpha = \mathbf{U}_{{}^w\alpha}$ where ${}^w\alpha(s) = \alpha({}^{n^{-1}}s) = \alpha(s)$. The quotient of $C_{\mathbf{G}}(s)$ by this subgroup is finite since it is isomorphic to a sub-quotient of $W(\mathbf{T})$. So this subgroup is indeed the identity component of $C_{\mathbf{G}}(s)$. ▪

We will often use for convenience the compact notation $C^\circ_{\mathbf{G}}$ for the connected component of a centralizer.

2.4 REMARK. The Weyl group $W^\circ(s)$ of $C^\circ_{\mathbf{G}}(s)$ is thus the group generated by the reflections s_α for which $\alpha(s) = 1$. It is a normal subgroup of the Weyl group of $C_{\mathbf{G}}(s)$ which is $W(s) = \{ w \in W(\mathbf{T}) \mid {}^w s = s \}$. The quotient $W(s)/W^\circ(s)$ is isomorphic to the quotient $C_{\mathbf{G}}(s)/C^\circ_{\mathbf{G}}(s)$.

2.5 PROPOSITION. *If $x = su$ is the Jordan decomposition of an element of $\mathbf{G}$, where s is semi-simple and u unipotent, then $x \in C^\circ_{\mathbf{G}}(s)$.*

PROOF: Let $\mathbf{B}$ be a Borel subgroup containing x, and $\mathbf{T}$ a maximal torus of $\mathbf{B}$ containing s. Let $\mathbf{B} = \mathbf{U}.\mathbf{T}$ be the corresponding Levi decomposition of $\mathbf{B}$ and write $u = \prod u_\alpha$ (with $u_\alpha \in \mathbf{U}_\alpha$ where $\mathbf{U} = \prod \mathbf{U}_\alpha$). Then for any root α such that $u_\alpha \neq 1$, we have $\alpha(s) = 1$ which implies that $\mathbf{U}_\alpha \subset C^\circ_{\mathbf{G}}(s)$, whence the result as $s \in \mathbf{T} \subset C^\circ_{\mathbf{G}}(s)$. ■

2.6 EXAMPLES.

(i) In the groups $\mathbf{GL}_n$ or $\mathbf{SL}_n$, centralizers of semi-simple elements are connected. Indeed such an element is conjugate to a diagonal matrix of the form $s = \operatorname{diag}(t_1, \ldots, t_n)$ where we may assume in addition that equal t_i are grouped in consecutive blocks, thereby defining a partition π of n. An easy computation then shows that the centralizer of s is the group of block-diagonal matrices defined by the partition π. Such a subgroup is easily seen to be connected; we will see in chapter 15 that it is a Levi subgroup. Using the description of chapter 15, it may also be argued that, if $\mathbf{T}$ is the torus consisting of diagonal matrices, then the elements of $W(\mathbf{T})$ (permutation matrices) which centralize s are products of generating reflections s_α which centralize s, *i.e.*, $W(s) = W^\circ(s)$ showing once again that the centralizer of s is connected.

(ii) We finish with an example of a semi-simple element whose centralizer is not connected. Consider the centralizer in $\mathbf{PGL}_2$ of $\begin{pmatrix} 1 & 0 \\ 0 & -1 \end{pmatrix}$; in characteristic different from 2, it has two connected components, consisting respectively of the matrices of the form $\begin{pmatrix} a & 0 \\ 0 & b \end{pmatrix}$ and of the matrices of the form $\begin{pmatrix} 0 & a \\ b & 0 \end{pmatrix}$.

References

The properties of the intersection of two parabolic subgroups are expounded in detail in [C. W. CURTIS, Reduction theorems for characters of finite groups of Lie type, *Journal of the mathematical society of Japan* **27** (1975), 666–688]. Numerous properties of parabolic subgroups are given in [BT]. A detailed study of the centralizers of semi-simple elements can be found in, *e.g.*, [D.I. DERIZIOTIS, Conjugacy classes and centralizers of semi-simple elements in finite groups of Lie type, *Vorlesungen aus dem Fachbereich Mathematik der Universität Essen* **11**, 1984].

3. RATIONALITY, THE FROBENIUS ENDOMORPHISM, THE LANG-STEINBERG THEOREM

We are interested in the groups of points over $\mathbb{F}_q$ of reductive algebraic groups over $\overline{\mathbb{F}}_q$ defined over $\mathbb{F}_q$. In this chapter, we recall basic facts about $\mathbb{F}_q$-structures on algebraic varieties and we give the Lang-Steinberg theorem along with some of its consequences.

3.1 DEFINITION. *An algebraic variety* $\mathbf{V}$ *over* $\overline{\mathbb{F}}_q$ *is* **defined over** $\mathbb{F}_q$, *or endowed with an* $\mathbb{F}_q$**-structure** $\mathbf{V}_\circ$, *if there exists a variety* $\mathbf{V}_\circ$ *over* $\mathbb{F}_q$ *such that* $\mathbf{V} = \mathbf{V}_\circ \otimes_{\mathbb{F}_q} \overline{\mathbb{F}}_q$. *The* **geometric Frobenius endomorphism** $F : \mathbf{V} \to \mathbf{V}$ *associated to this* $\mathbb{F}_q$*-structure is then defined as the endomorphism* $F_\circ \otimes \mathrm{Id}$ *where* $F_\circ$ *is the endomorphism of* $\mathbf{V}_\circ$ *that raises the functions on* $\mathbf{V}_\circ$ *to the* q*-th power. The endomorphism* Φ *of* $\mathbf{V}$ *induced by the element* $\lambda \mapsto \lambda^q$ *of* $\mathrm{Gal}(\overline{\mathbb{F}}_q/\mathbb{F}_q)$ *is called the* **arithmetic Frobenius endomorphism**.

We now explain the above definition in terms of rings of functions in the case of an affine or projective variety.

Recall that an affine (resp. projective) variety over a field K is defined by a finitely generated K-algebra (resp. graded reduced K-algebra generated by its elements of degree one). A closed subvariety is defined by an ideal (resp. a homogeneous ideal). If the variety $\mathbf{V}$ is affine or projective, it is defined over $\mathbb{F}_q$ if and only if the corresponding $\overline{\mathbb{F}}_q$-algebra can be written as $A = A_\circ \otimes_{\mathbb{F}_q} \overline{\mathbb{F}}_q$ where $A_\circ$ is a finitely generated $\mathbb{F}_q$-algebra. We say that $A_\circ$ is an $\mathbb{F}_q$-structure on A (or on $\mathbf{V}$). The Frobenius endomorphism $F : \mathbf{V} \to \mathbf{V}$ associated to this $\mathbb{F}_q$-structure is then defined by the endomorphism of $A = A_\circ \otimes_{\mathbb{F}_q} \overline{\mathbb{F}}_q$ given by $a \otimes \lambda \mapsto a^q \otimes \lambda$ (in a coordinate system for the variety, the Frobenius morphism raises each coordinate to the q-th power). The arithmetic Frobenius endomorphism Φ maps $a \otimes \lambda$ to $a \otimes \lambda^q$. The composite map $F \circ \Phi$ raises each element of A to the q-th power, so is the identity on points of $\mathbf{V}$ over $\overline{\mathbb{F}}_q$.

3.2 EXAMPLE. Take for $\mathbf{V}$ the affine line over $\overline{\mathbb{F}}_q$: it is the affine variety defined by the $\overline{\mathbb{F}}_q$-algebra $\overline{\mathbb{F}}_q[T]$. The affine line $\mathbf{V}_\circ$ on $\mathbb{F}_q$ (defined by the $\mathbb{F}_q$-algebra $\mathbb{F}_q[T]$) is an $\mathbb{F}_q$-structure on $\mathbf{V}$ as we have $\overline{\mathbb{F}}_q[T] = \mathbb{F}_q[T] \otimes_{\mathbb{F}_q} \overline{\mathbb{F}}_q$. The geometric Frobenius endomorphism maps a polynomial $P(T)$ to $P(T^q)$, while the arithmetic Frobenius morphism maps $P(T) = \sum_i a_i T^i$ to $\sum_i a_i^q T^i$. So $F \circ \Phi$ maps $P(T)$ to $P(T)^q$. A point of $\mathbf{V}$ over $\overline{\mathbb{F}}_q$ corresponds to an element $a \in \overline{\mathbb{F}}_q$: it is defined by the kernel of the morphism $P \mapsto P(a) : \overline{\mathbb{F}}_q[T] \to \overline{\mathbb{F}}_q$; the image of this point by $F \circ \Phi$ is the point defined by the kernel of $P \mapsto P(a)^q$ which is the same, so $F \circ \Phi$ is the identity on points over $\overline{\mathbb{F}}_q$.

Note that whereas F is a morphism of $\overline{\mathbb{F}}_q$-varieties, the arithmetic Frobenius endomorphism is only a morphism of $\mathbb{F}_q$-varieties. We shall often forget the word "geometric" and call F the Frobenius endomorphism. A morphism (resp. subvariety, *etc.*) is said to be **rational**, or defined over $\mathbb{F}_q$ if it is stable under the action of the Frobenius endomorphism. The following proposition gives basic results about $\mathbb{F}_q$-structures on affine or projective varieties.

3.3 PROPOSITION. *Let* $\mathbf{V}$ *be an affine or projective variety over* $\overline{\mathbb{F}}_q$ *and let* A *be its algebra.*

(i) *Let* F *be a surjective morphism of* $\overline{\mathbb{F}}_q$*-algebras from* A *to* A^q; *then* F *is the Frobenius endomorphism associated to an* $\mathbb{F}_q$*-structure over* $\mathbf{V}$ *if and only if for any* $x \in A$ *there exists a positive integer* n *such that* $F^n(x) = x^{q^n}$.

In the next two items, we assume that $\mathbf{V}$ *has an* $\mathbb{F}_q$*-structure* A_0 *and that* F *is the associated Frobenius endomorphism.*

(ii) *We have* $A_0 = \{ x \in A \mid F(x) = x^q \}$.

(iii) *A closed subvariety of* $\mathbf{V}$ *is* F*-stable if and only if its ideal is of the form* $I_0 \otimes \overline{\mathbb{F}}_q$ *where* I_0 *is an ideal of* A_0. *In this case the subvariety is defined over* $\mathbb{F}_q$ *as a variety and its Frobenius endomorphism is the restriction of* F.

PROOF: First we recall the following result.

3.4 PROPOSITION. *Let* L *be a finite Galois extension of a field* K, *and let* V *be an* L*-vector space; we assume that there exists a* K*-linear action of the group* $\mathrm{Gal}(L/K)$ *on* V *such that* $\sigma(\lambda v) = \sigma(\lambda)\sigma(v)$ *for any* $\sigma \in \mathrm{Gal}(L/K)$, $\lambda \in L$ *and* $v \in V$. *Then the* K*-subspace* V° *of* V *consisting of the fixed points under* $\mathrm{Gal}(L/K)$ *defines a* K*-structure on* V, *i.e., we have* $V = V^\circ \otimes_K L$.

PROOF: See Bourbaki *Algèbre* chap. 5, 10.4, proposition 7. ■

This proposition is helpful because of the following corollary.

3.5 COROLLARY. *Let* V *be an* $\overline{\mathbb{F}}_q$*-vector space and let* Φ *be an* $\mathbb{F}_q$*-linear endomorphism of* V *such that* $\Phi(\lambda v) = \lambda^q \Phi(v)$ *for any* $v \in V$ *and* $\lambda \in \overline{\mathbb{F}}_q$; *assume that for each* $x \in V$ *there exists an integer* $n > 0$ *such that* $\Phi^n(x) = x$. *Then* $V = V^\Phi \otimes_{\mathbb{F}_q} \overline{\mathbb{F}}_q$ *and in this decomposition* $\Phi(a \otimes \lambda) = a \otimes \lambda^q$.

Note that, if V is the algebra of an $\overline{\mathbf{F}}_q$-variety, the conclusion of the corollary is that V^Φ is an $\mathbf{F}_q$-structure on V and that Φ is the associated arithmetic Frobenius endomorphism.

PROOF: Let $x \in V$, let $n > 0$ be such that $\Phi^n(x) = x$ and let V_x be the $\mathbf{F}_{q^n}$-subspace of V generated by $x, \Phi(x), \ldots, \Phi^{n-1}(x)$. We see that Φ^n acts trivially on V_x, so, by 3.4 applied with $L = \mathbf{F}_{q^n}$ and $K = \mathbf{F}_q$, we get $V_x = V_x^\Phi \otimes_{\mathbf{F}_q} \mathbf{F}_{q^n}$, so that $x \in V_x^\Phi \otimes_{\mathbf{F}_q} \mathbf{F}_{q^n}$, and in particular $x \in V^\Phi \otimes_{\mathbf{F}_q} \overline{\mathbf{F}}_q$. Moreover, for any $a \in V^\Phi$ and $\lambda \in \overline{\mathbf{F}}_q$ we have

$$\Phi(a \otimes \lambda) = \lambda^q \Phi(a) = \lambda^q a = a \otimes \lambda^q.$$

■

We can now prove proposition 3.3. First, we prove (i). If F is the Frobenius endomorphism associated to an $\mathbf{F}_q$-structure on $\mathbf{V}$ and if $x = \sum_i x_i \otimes \lambda_i$ with $x_i \in A_\circ$, $\lambda_i \in \overline{\mathbf{F}}_q$, then we have $x^{q^n} = \sum_i x_i^{q^n} \otimes \lambda_i^{q^n}$ whence $x^{q^n} = F^n(x)$ if n is such that all λ_i are in $\mathbf{F}_{q^n}$.

Conversely, if for any $x \in A$ we have $x^{q^n} = F^n(x)$ for some n, consider the morphism Φ of $\mathbf{F}_q$-algebras defined by $\Phi(x) = F^{-1}(x^q)$ (this is possible because the assumption on F and the injectivity of $x \mapsto x^q$ imply that F is injective). Then Φ satisfies the assumptions of 3.5 (with $V = A$); so $A_\circ = A^\Phi$ is an $\mathbf{F}_q$-structure on $\mathbf{V}$. The Frobenius endomorphism associated to this $\mathbf{F}_q$-structure is defined by $a \otimes \lambda \mapsto a^q \otimes \lambda$; but we have $F(a \otimes \lambda) = F(a) \otimes \lambda = a^q \otimes \lambda$, the former equality because F is an algebra morphism, the latter because we have $\Phi(x) = x$ if and only if $F(x) = x^q$, so $A_\circ = \{ x \in A \mid F(x) = x^q \}$. Thus we see that F is the Frobenius endomorphism associated to the $\mathbf{F}_q$-structure A^Φ, whence (i). We have also got (ii).

Note that our argument proves that giving an $\mathbf{F}_q$-structure, a Frobenius endomorphism F and an arithmetic Frobenius endomorphism Φ are equivalent, the correspondence between F and Φ being given by the formula $\Phi(x) = F^{-1}(x^q)$.

Let us prove (iii). If a closed subvariety is F-stable the ideal I of A by which it is defined is F-stable. If I is F-stable F is surjective from I to I^q: indeed for any $y \in I^q$ there exists $x \in A$ such that $F(x) = y$, and there exists a positive integer n such that $F^n(x) = x^{q^n}$; as $F^n(x) = F^{n-1}(y)$, we have $x^{q^n} \in I$ because I is F-stable, so x is in I as I is equal to its root. We now notice that the proof of (i) and (ii) is valid for an ideal, as it does not use the fact that A is unitary, so I is equal to $I_\circ \otimes_{\mathbf{F}_q} \overline{\mathbf{F}}_q$ with $I_\circ = I \cap A_\circ$. The converse and the other properties are clear. ■

Note that the property (iii) above shows that a closed rational subvariety is defined over $\mathbf{F}_q$ as a variety. This result can be extended to any subvariety, so that the terminology is consistent.

We now give a second list of basic results on $\mathbf{F}_q$-structures dealing mostly with Frobenius endomorphisms. We shall give the proofs only for affine varieties or projective varieties, but they can be easily extended to quasi-projective varieties. All the varieties that we shall consider in this book will be quasi-projective.

3.6 PROPOSITION. *Let* $\mathbf{V}$ *be an algebraic variety over* $\overline{\mathbf{F}}_q$ *endowed with an* $\mathbf{F}_q$*-structure with corresponding Frobenius endomorphism* F.

(i) *Let* φ *be an automorphism of* $\mathbf{V}$ *such that* $(\varphi F)^n = F^n$ *for some positive integer* n*; then* φF *is the Frobenius endomorphism associated to some* $\mathbf{F}_q$*-structure over* $\mathbf{V}$.

(ii) *If* F' *is another Frobenius endomorphism corresponding to an* $\mathbf{F}_q$*-structure over* $\mathbf{V}$*, there exists a positive integer* n *such that* $F^n = F'^n$.

(iii) F^n *is the Frobenius endomorphism associated to some* $\mathbf{F}_{q^n}$*-structure over* $\mathbf{V}$.

(iv) *Any closed subvariety of a variety defined over* $\mathbf{F}_q$ *is defined over a finite extension of* $\mathbf{F}_q$*. Any morphism from a variety defined over* $\mathbf{F}_q$ *to another one is defined over a finite extension of* $\mathbf{F}_q$.

(v) *The* F*-orbits in the set of points of* $\mathbf{V}$ *and the set* $\mathbf{V}^F$ *of rational points of* $\mathbf{V}$ *(also denoted by* $\mathbf{V}(\mathbf{F}_q)$*) are finite.*

PROOF: As before, we denote by $A = A_\circ \otimes_{\mathbf{F}_q} \overline{\mathbf{F}}_q$ the algebra of $\mathbf{V}$. Let us prove (i). As φ is bijective, the endomorphism φF is surjective on A^q. It is also clear that it satisfies the second condition of 3.3 (i), whence the result.

As A is finitely generated, there exists a positive integer n such that $F'^n(x) = F^n(x) = x^{q^n}$ for all $x \in A$, whence (ii).

We prove (iii). Since tensor products are associative we have

$$A = A_\circ \otimes_{\mathbf{F}_q} \mathbf{F}_{q^n} \otimes_{\mathbf{F}_{q^n}} \overline{\mathbf{F}}_q.$$

Let $B_\circ = A_\circ \otimes_{\mathbf{F}_q} \mathbf{F}_{q^n}$; the endomorphism F^n maps the element $b \otimes \lambda$ of $B \otimes_{\mathbf{F}_{q^n}} \overline{\mathbf{F}}_q$ to $b^{q^n} \otimes \lambda$, so that B defines an $\mathbf{F}_{q^n}$-structure on $\mathbf{V}$ for which the Frobenius endomorphism is F^n, which is the desired result.

The ideal defining a closed subvariety is of finite type so, by 3.3 (i), by (iii) above and by 3.3 (iii), we get the first assertion of (iv). The same kind of argument applies to the second assertion.

Let us prove (v). A point X of $\mathbf{V}$ is a morphism $x : A \to \overline{\mathbb{F}}_q$. If $\{a_0, \ldots, a_n\}$ is a set of generators for A_0, the point X is fixed by F^n for any n such that all $x(a_i)$ are in $\mathbb{F}_{q^n}$, so the orbit of X under F is finite. A rational point (over $\mathbb{F}_q$) corresponds to a morphism of the form $x_0 \otimes 1$, *i.e.*, to an algebra morphism $x_0 : A_0 \to \mathbb{F}_q$. There are only a finite number of such morphisms, as there is only a finite set of possible images for the a_i. ■

3.7 EXAMPLE. Let $\mathbf{V} \simeq \mathbb{A}^n$ be an affine space of dimension n on $\overline{\mathbb{F}}_q$; then $|\mathbf{V}^F| = q^n$ for any $\mathbb{F}_q$-structure on $\mathbf{V}$.

This fact will be proved as an immediate application of the properties of l-adic cohomology (see 10.11 (ii)).

3.8 EXERCISE. Show that any $\mathbb{F}_q$-structure on the affine line $\mathbb{A}^1$ is given by a Frobenius endomorphism on $\overline{\mathbb{F}}_q[T]$ of the form $T \mapsto aT^q$ with $a \in \overline{\mathbb{F}}_q^{\times}$.

An algebraic group over $\overline{\mathbb{F}}_q$ is said to be **defined over** $\mathbb{F}_q$ if it has an $\mathbb{F}_q$-structure such that the Frobenius endomorphism is a group morphism. The topic of this book is the study of finite groups arising as groups of rational points of reductive groups over $\overline{\mathbb{F}}_q$ defined over $\mathbb{F}_q$. They are called **finite groups of Lie type**.

REMARK. For the sake of simplicity we exclude from this book Ree and Suzuki groups. These groups arise as groups of fixed points in an algebraic group under an endomorphism which is not a Frobenius endomorphism as defined above, but whose square or cube is a Frobenius endomorphism. The classification theorem 3.17 below can be extended to such situations. Suzuki and Ree groups arise from root systems of type B_2, G_2, or F_4.

3.9 EXAMPLE(S). Let us consider the group $\mathbf{GL}_n$ over $\overline{\mathbb{F}}_q$. It is defined over $\mathbb{F}_q$ as its algebra is $\overline{\mathbb{F}}_q[T_{i,j}, \det(T_{i,j})^{-1}]$, which is isomorphic to

$$\mathbb{F}_q[T_{i,j}, \det(T_{i,j})^{-1}] \otimes \overline{\mathbb{F}}_q.$$

Its points over $\mathbb{F}_q$ form the group $\mathbf{GL}_n(\mathbb{F}_q)$. The analogous statements are clearly true for the special linear, the symplectic, the orthogonal, *etc.* groups (see chapter 15 for more details). Any embedding of an algebraic group $\mathbf{G}$ into $\mathbf{GL}_n$ as above defines a **standard** Frobenius endomorphism on $\mathbf{G}$ by restriction of the endomorphism of $\mathbf{GL}_n$ defined by $T_{ij} \mapsto T_{ij}^q$. But there are other examples of rational structures on algebraic groups; for instance the unitary group is $\mathbf{GL}_n^{F'}$ where F' is the Frobenius endomorphism defined by $F'(x) = F({}^t x^{-1})$, with F being the standard Frobenius endomorphism on $\mathbf{GL}_n$. We shall see (chapter 15) that F' is not standard.

The following theorem is of paramount importance in the theory of algebraic groups over finite fields.

3.10 THE LANG-STEINBERG THEOREM. *Let* $\mathbf{G}$ *be a connected affine algebraic group and let* F *be a surjective endomorphism of* $\mathbf{G}$ *with a finite number of fixed points; then the map* $\mathcal{L} : g \mapsto g^{-1}.{}^F g$ *from* $\mathbf{G}$ *to itself is surjective.*

OUTLINE OF PROOF (SEE [ST1, 10]): First one proves that the differential at 1 of such an endomorphism is a nilpotent map. From this one deduces easily that the differential of $\mathcal{L}$ at 1 is surjective, which implies that $\mathcal{L}$ is a dominant morphism of algebraic varieties, *i.e.*, that its image contains a dense open subset. Then one shows that for any x the map $g \mapsto g^{-1}.x.{}^F g$ also has a surjective differential at 1, so its image also contains a dense open subset. By the connectedness (whence the irreducibility) of $\mathbf{G}$, these two dense open subsets must have a non-empty intersection; so there exists g and h such that $g^{-1}.{}^F g = h^{-1}.x.{}^F h$, whence $x = \mathcal{L}(gh^{-1})$. ∎

By 3.6 (v) and the remarks before 3.2, a Frobenius endomorphism satisfies the assumptions of 3.10.

3.11 DEFINITION. *The map* $\mathcal{L}$ *of the above theorem will be called* **the Lang map**.

The following corollary shows the importance of the Lang-Steinberg theorem.

3.12 COROLLARY. *Let* $\mathbf{V}$ *be an algebraic variety defined over* $\mathbb{F}_q$*, and assume that an algebraic connected group* $\mathbf{G}$ *defined over* $\mathbb{F}_q$ *acts on* $\mathbf{V}$ *by an action defined over* $\mathbb{F}_q$*. Then any* F*-stable* $\mathbf{G}$*-orbit contains a rational point.*

PROOF: Let v be a point of an F-stable orbit, so that we have ${}^F v = {}^g v$ for some element $g \in \mathbf{G}$. By the Lang-Steinberg theorem the element g can be written $h^{-1F}h$ with $h \in \mathbf{G}$. So ${}^{F_h F}v = {}^h v$, *i.e.*, ${}^F({}^h v) = {}^h v$, and ${}^h v$ is an F-fixed point in the orbit of v. ∎

3.13 COROLLARY. *If* $\mathbf{H}$ *is a closed connected rational subgroup of the algebraic group* $\mathbf{G}$ *(defined over* $\mathbb{F}_q$*), then* $(\mathbf{G}/\mathbf{H})^F = \mathbf{G}^F/\mathbf{H}^F$.

PROOF: By the above corollary any rational left $\mathbf{H}$-coset contains a rational point. So the natural map $\mathbf{G}^F/\mathbf{H}^F \to (\mathbf{G}/\mathbf{H})^F$ is surjective. It is injective since, if $x, y \in \mathbf{G}^F$ are in the same $\mathbf{H}$-coset, then $x^{-1}y$ is in $\mathbf{H}^F$. ∎

3.14 EXAMPLE. *A trap.* Let us consider the group $\mathbf{PSL}_n$ over $\overline{\mathbb{F}}_q$, *i.e.*, the quotient group of $\mathbf{SL}_n$ by its centre. This group is defined over $\mathbb{F}_q$ (for the standard Frobenius endomorphism; see 3.9), but if n is not relatively prime to $q-1$ the group $\mathbf{PSL}_n^F$ is not the quotient of $\mathbf{SL}_n(\mathbb{F}_q)$ by its centre (we know by 3.13 that this kind of phenomenon happens only in the case of a quotient by a non-connected group). Indeed, the centre μ_n of $\mathbf{SL}_n$ consists of the scalar matrices which are equal to an n-th root of unity times the identity. The image in $\mathbf{PSL}_n$ of $x \in \mathbf{SL}_n$ is in $\mathbf{PSL}_n^F$ if and only if $x.{}^F x^{-1} \in \mu_n$. If n is relatively prime to $q-1$, the map $z \mapsto z.{}^F z^{-1} = z^{1-q}$ is bijective from μ_n onto μ_n and, as in the proof of 3.12, we deduce that x is equal to an element of $\mathbf{SL}_n^F$ up to multiplication by an element of μ_n yielding $\mathbf{PSL}_n^F = \mathbf{SL}_n^F/\mu_n^F$; note that for such an n the centre μ_n^F of $\mathbf{SL}_n^F$ is the identity. For other values of n the group $\mathbf{PSL}_n^F$ contains elements that are not in $\mathbf{SL}_n^F/\mu_n^F$; this can be seen, *e.g.*, by making $\mathbf{SL}_n$ act by translation on $\mathbf{SL}_n/\mu_n$ and by applying the results of 3.21 below (whence a bijection from $(\mathbf{SL}_n/\mu_n)^F$ onto the set of F-conjugacy classes of μ_n).

We give below other important (and easy to prove) consequences of 3.12.

3.15 APPLICATIONS. Let $\mathbf{G}$ be an algebraic group defined over $\mathbb{F}_q$.

(i) In $\mathbf{G}$, there exist rational Borel subgroups and any two of them are conjugate under $\mathbf{G}^F$.

(ii) In any rational Borel subgroup, there exists a rational maximal torus.

These two properties come from the fact that Borel subgroups (resp. maximal tori in a Borel subgroup) are one orbit under the connected component $\mathbf{G}^\circ$ of $\mathbf{G}$ (resp. under the Borel subgroup).

(iii) Any rational parabolic subgroup $\mathbf{P}$ has a rational Levi decomposition and two rational Levi subgroups are conjugate by a rational element of the unipotent radical of $\mathbf{P}$ (see 1.18).

Indeed two Levi subgroups are conjugate by an element v of the unipotent radical of $\mathbf{P}$. But, if $\mathbf{L}$ and ${}^v\mathbf{L}$ are both rational, then the element $v^{-1\,F}v$ normalizes $\mathbf{L}$ and is in the unipotent radical of $\mathbf{P}$, so is 1.

By (i), (ii), (iii) we see that rational maximal tori contained in rational Borel subgroups are conjugate under $\mathbf{G}^F$.

(iv) Any rational conjugacy class of $\mathbf{G}$ contains a rational element.

(v) Let $\mathbf{T}$ be a rational maximal torus of $\mathbf{G}$; the Frobenius endomorphism F acts on the Weyl group W of $\mathbf{T}$, and we have $W^F = N_{\mathbf{G}}(\mathbf{T})^F/\mathbf{T}^F$.

(vi) Let $\mathbf{T}$ and $\mathbf{B} \supset \mathbf{T}$ be respectively a rational maximal torus and a rational Borel subgroup of $\mathbf{G}$; then $\mathbf{G} = \coprod_w \mathbf{B}w\mathbf{B}$, which gives, as $\mathbf{B}$ is connected, the following "rational Bruhat decomposition" $\mathbf{G}^F =$

$\coprod_{w \in W^F} \mathbf{B}^F w \mathbf{B}^F$ (it is the Bruhat decomposition associated to the (B, N) -pair of $\mathbf{G}^F$ to which we have already referred; see 1.3).

3.16 COROLLARY. *Let* $\mathbf{G}$ *be a connected algebraic group defined over* $\mathbb{F}_q$*; then any rational semi-simple element lies in a rational maximal torus of* $\mathbf{G}$.

PROOF: Let s be a rational semi-simple element; as any semi-simple element lies in a maximal torus of $\mathbf{G}$ (see 0.9 (i)), there is such a torus containing s, and this torus is contained in the connected algebraic group $C^\circ_{\mathbf{G}}(s)$. As s is central in its centralizer, it is in any maximal torus of $C^\circ_{\mathbf{G}}(s)$, and in particular in the rational maximal tori which exist by 3.15 and are also rational maximal tori of $\mathbf{G}$. ∎

If $\mathbf{T}$ is a torus defined over $\mathbb{F}_q$, the algebra of the affine variety $\mathbf{T}$ is of the form $A(\mathbf{T}) = A_\circ(\mathbf{T}) \otimes_{\mathbb{F}_q} \overline{\mathbb{F}}_q$, and we have

$$X(\mathbf{T}) = \mathrm{Hom}(\mathbf{T}, \mathbb{G}_m) = \mathrm{Hom}(\overline{\mathbb{F}}_q[T, T^{-1}], A_\circ(\mathbf{T}) \otimes_{\mathbb{F}_q} \overline{\mathbb{F}}_q).$$

The Galois group $\mathrm{Gal}(\overline{\mathbb{F}}_q/\mathbb{F}_q)$ acts on the algebras of $\mathbb{G}_m$ and of $\mathbf{T}$ (the element $x \mapsto x^q$ acts on both algebras by the arithmetic Frobenius endomorphism). This defines an action of $\mathrm{Gal}(\overline{\mathbb{F}}_q/\mathbb{F}_q)$ on $X(\mathbf{T})$. Let τ be the action on $X(\mathbf{T})$ of the element $x \mapsto x^q$ of $\mathrm{Gal}(\overline{\mathbb{F}}_q/\mathbb{F}_q)$. For $\alpha \in X(\mathbf{T})$ and $t \in \mathbf{T}$ the following formula is easily checked, using the fact that the composition of the arithmetic and the geometric Frobenius endomorphisms is the identity on points in $\overline{\mathbb{F}}_q$ (see the remarks after 3.1):

$$(\tau\alpha)({}^F t) = (\alpha(t))^q.$$

If $\mathbf{T}$ is a rational maximal torus of a reductive connected algebraic group $\mathbf{G}$, it is clear that τ permutes the root system of $\mathbf{G}$ relative to $\mathbf{T}$. Similarly the transpose automorphism τ^* of $Y(\mathbf{T})$ permutes the coroots. With these notations one can give the following classification theorem, analogous to 0.45.

3.17 THEOREM. *The datum* $(X(\mathbf{T}), Y(\mathbf{T}), \Phi, \check{\Phi})$ *together with the action of* τ *on* $X(\mathbf{T})$ *characterizes the pair* $(\mathbf{G}, F)$ *up to isomorphism; and for any root datum, any* q *and any automorphism* τ *of* $X(\mathbf{T})$ *such that* τ *permutes the roots and* τ^* *permutes the coroots, there exists a pair* $(\mathbf{G}, F)$ *as above with this datum.*

PROOF: See [Sp, 11.14.9]. ∎

In the above theorem τ induces an automorphism of the root system of $\mathbf{G}$ relative to $\mathbf{T}$. If $\mathbf{T}$ is in a rational Borel subgroup, this automorphism stabilizes a basis of the root system. The list of non-trivial such automorphisms of irreducible root systems is

$$^2A_n, \quad ^2D_n, \quad ^3D_4, \quad ^2E_6$$

(see [Bbk, VI, §4]), where the root systems are named as after 0.43, and a left exponent denotes the order of the automorphism.

We give now some "finite group theoretic" results, in particular about characterizations of semi-simple and unipotent elements.

3.18 PROPOSITION. *Let* $\mathbf{G}$ *be an algebraic group over* $\overline{\mathbb{F}}_q$*. Then the semi-simple elements of* $\mathbf{G}$ *are the* p'*-elements of* $\mathbf{G}$ *and the unipotent elements are the* p*-elements of* $\mathbf{G}$*, if* p *is the characteristic of* $\mathbb{F}_q$*.*

PROOF: This result is true in $\mathbf{GL}_n$, so is true in any linear algebraic group (by definition a subgroup of some $\mathbf{GL}_n$). ∎

3.19 PROPOSITION.

(i) *Let* $\mathbf{G}$ *be a connected algebraic group defined over* $\mathbb{F}_q$*; then the Sylow* p*-subgroups of* $\mathbf{G}^F$ *are the groups* $\mathbf{U}^F$ *where* $\mathbf{U}$ *runs over the set of the unipotent radicals of the rational Borel subgroups of* $\mathbf{G}$*.*

(ii) *If* $\mathbf{G}$ *is reductive, we have* $|\mathbf{G}^F| = q^{|\Phi^+|}|\mathbf{T}^F|(\sum_{w\in W^F} q^{l(w)})$ *where* $\mathbf{T}$ *is a rational maximal torus contained in a rational Borel subgroup* $\mathbf{B}$ *and where* Φ^+ *is the set of roots of* $\mathbf{G}$ *relative to* $\mathbf{T}$ *which are positive for the order defined by* $\mathbf{B}$*.*

PROOF: If $\mathbf{G}$ is not reductive we consider its quotient by its unipotent radical which is a p-subgroup by 3.18. This is contained in the unipotent radical of any Borel subgroup by 0.15 and is connected so that we have by 3.13 $|\mathbf{G}^F| = |(\mathbf{G}/R_u(\mathbf{G}))^F||R_u(\mathbf{G})^F|$.

So we may assume that $\mathbf{G}$ is reductive. Let us first prove (ii). Let $\mathbf{U}$ be the unipotent radical of $\mathbf{B}$; the Bruhat decomposition, as stated in 1.7, implies

$$|\mathbf{G}^F/\mathbf{B}^F| = \sum_{w\in W^F} |\mathbf{U}_w^F|$$

(see definition of $\mathbf{U}_w$ below 1.6). As $\mathbf{U}_w^F$ is, by 0.29, an affine space of dimension $l(w)$, by 3.7 we have $|\mathbf{U}_w^F| = q^{l(w)}$, whence (ii), as we have, also by 3.7, $|\mathbf{B}^F| = |\mathbf{T}^F||\mathbf{U}^F|$ and $|\mathbf{U}^F| = q^{|\Phi^+|}$.

Property (i) is then a consequence of the fact that

$$|\mathbf{G}^F/\mathbf{U}^F| = |\mathbf{T}^F|(\sum_{w\in W^F} q^{l(w)})$$

is relatively prime to p, which is clear as $\mathbf{T}$ is a p'-group and $\sum_{w\in W^F} q^{l(w)} \equiv 1 \pmod q$. (It can be proved that $N_{\mathbf{G}^F}(\mathbf{U}^F) = \mathbf{B}^F$ so that the integer $\sum_{w\in W^F} q^{l(w)}$ is actually equal to the number of p-Sylow subgroups of $\mathbf{G}^F$; see, *e.g.*, [BT, 5.19].) ∎

3.20 COROLLARY. *Let* $\mathbf{G}$ *be a connected algebraic group defined over* $\mathbb{F}_q$*; any rational unipotent element of* $\mathbf{G}$ *lies in a rational Borel subgroup.*

PROOF: Any rational unipotent element is in a Sylow p-subgroup by 3.18, so by 3.19 is in the unipotent radical of a rational Borel subgroup. ∎

We now give another example of application of the Lang-Steinberg theorem, which is a refinement of 3.12 and shows how the set of rational points of an F-stable $\mathbf{G}$-orbit splits into $\mathbf{G}^F$-orbits.

3.21 PROPOSITION. *Make the same assumptions as in 3.12; let* $\mathcal{O}$ *be an* F*-stable orbit in* $\mathbf{V}$ *and let* x *be an element of* $\mathcal{O}^F$ *(such an* x *exists by 3.12).*

(i) *Let* $g \in \mathbf{G}$*; then* $gx \in \mathcal{O}^F$ *if and only if* $g^{-1}{}^F g \in \operatorname{Stab}_{\mathbf{G}}(x)$.

(ii) *There is a well-defined map which sends the* $\mathbf{G}^F$*-orbit of* $gx \in \mathcal{O}^F$ *to the* F*-conjugacy class of the image of* $g^{-1}{}^F g$ *in* $\operatorname{Stab}_{\mathbf{G}}(x)/\operatorname{Stab}_{\mathbf{G}}(x)^\circ$ *and it is a bijection.*

PROOF: First recall that F-**conjugation** in a group K means the action of K on itself defined for any $k \in K$ by $x \mapsto kx({}^F k)^{-1}$. Property (i) comes from an easy computation. Before proving (ii), let us note that $hx = gx$ if and only if h and g differ by an element of $\operatorname{Stab}_{\mathbf{G}}(x)$, and in this case the elements $h^{-1}{}^F h$ and $g^{-1}{}^F g$ are F-conjugate in $\operatorname{Stab}_{\mathbf{G}}(x)$. So we get a map from $\mathcal{O}^F$ to the set of F-conjugacy classes of $\operatorname{Stab}_{\mathbf{G}}(x)$. But if h is in $\mathbf{G}^F$, the elements gx and hgx have the same image $g^{-1}{}^F g = (hg)^{-1}{}^F(hg)$. So we get a map from the set of $\mathbf{G}^F$-orbits in $\mathcal{O}^F$ to the set of F-conjugacy classes in the group $\operatorname{Stab}_{\mathbf{G}}(x)$. Let us prove injectivity: if $g^{-1}{}^F g$ and $h^{-1}{}^F h$ are F-conjugate by an element $n \in \operatorname{Stab}_{\mathbf{G}}(x)$, then gnh^{-1} is in $\mathbf{G}^F$ and maps hx to gx, so hx and gx are in the same orbit. The surjectivity results from the Lang-Steinberg theorem as any element of $\operatorname{Stab}_{\mathbf{G}}(x)$ can be written $g^{-1}{}^F g$ with $g \in \mathbf{G}$. We are left with proving that the quotient map $\operatorname{Stab}_{\mathbf{G}}(x) \to \operatorname{Stab}_{\mathbf{G}}(x)/\operatorname{Stab}_{\mathbf{G}}(x)^\circ$ induces a bijection on F-conjugacy classes, which is a consequence of the following lemma.

3.22 LEMMA. *Let* $\mathbf{H}$ *be an algebraic group defined over* $\mathbb{F}_q$*, and let* $\mathbf{K}$ *be a closed connected normal subgroup defined over* $\mathbb{F}_q$*; then the quotient map induces a bijection from the set of* F*-conjugacy classes in* $\mathbf{H}$ *to that of* F*-conjugacy classes in* $\mathbf{H}/\mathbf{K}$.

PROOF: It is clear that F-conjugate elements are mapped to F-conjugate elements. Conversely, if h and h' are F-conjugate modulo $\mathbf{K}$, we have $kh = xh'{}^F x^{-1}$, with $x \in \mathbf{H}$ and $k \in \mathbf{K}$. As $\mathbf{K}$ is connected, we can apply in $\mathbf{K}$ the Lang-Steinberg theorem with the Frobenius endomorphism $k \mapsto {}^{hF}k$ (see 3.6 (i)). So k can be written $y^{-1}\,{}^{hF}y$, hence $kh = y^{-1}h\,{}^F y$ and $h = (yx)h'{}^F(yx)^{-1}$ and thus h and h' are F-conjugate elements of $\mathbf{H}$. ∎

3.23 APPLICATION. *Let $\mathbf{T}$ be a given rational maximal torus of a reductive connected algebraic group $\mathbf{G}$ defined over $\mathbb{F}_q$. The $\mathbf{G}^F$-conjugacy classes of rational maximal tori of $\mathbf{G}$ are parametrized by the F-conjugacy classes of $N_{\mathbf{G}}(\mathbf{T})/N_{\mathbf{G}}(\mathbf{T})^\circ = W(\mathbf{T})$.*

PROOF: We just apply proposition 3.21, taking for $\mathbf{V}$ the set of maximal tori of $\mathbf{G}$, on which $\mathbf{G}$ acts by conjugation. ∎

3.24 DEFINITION. *The F-conjugacy class of the image of $g^{-1}\,{}^F g$ in $W(\mathbf{T})$ is called* **the type** *of the torus ${}^g\mathbf{T}$ with respect to $\mathbf{T}$; a representative of this F-conjugacy class will be called* **a type** *of ${}^g\mathbf{T}$ with respect to $\mathbf{T}$.*

Let us note that by the conjugation action of g^{-1}, the torus ${}^g\mathbf{T}$, endowed with the action of F, is identified with the torus $\mathbf{T}$ endowed with the action of wF, if w is the image of $g^{-1}\,{}^F g$ in $W(\mathbf{T})$.

3.25 ANOTHER APPLICATION. *The $\mathbf{G}^F$-conjugacy classes of rational elements conjugate to some fixed $x \in \mathbf{G}^F$ under $\mathbf{G}$ are parametrized by the F-conjugacy classes of $C_{\mathbf{G}}(x)/C_{\mathbf{G}}(x)^\circ$.*

PROOF: We now apply 3.21 with $\mathbf{V} = \mathbf{G}$, the group $\mathbf{G}$ acting by conjugation. ∎

Two elements of $\mathbf{G}^F$ which are conjugate under $\mathbf{G}$ are said to be **geometrically conjugate**.

3.26 EXAMPLES.

(i) One can see that in $\mathbf{GL}_n$ centralizers of all elements are connected (we saw this in the previous chapter for semi-simple elements; see 2.6).

(ii) Two elements of $\mathbf{SL}_n(\overline{\mathbb{F}}_q)$ which are conjugate in $\mathbf{GL}_n(\overline{\mathbb{F}}_q)$ are conjugate in $\mathbf{SL}_n(\overline{\mathbb{F}}_q)$. Indeed, if $x = yx'y^{-1}$, then we also have $x = y'x'y'^{-1}$ with $y' = (\det y)^{-1/n}y \in \mathbf{SL}_n$. But x and x' are not necessarily conjugate under $\mathbf{SL}_n^F$, even if they are rational (here we choose F to be the standard Frobenius endomorphism; see 3.9). Let us consider, for instance the unipotent element $u = \begin{pmatrix} 1 & 1 \\ 0 & 1 \end{pmatrix}$ of $\mathbf{SL}_2^F$. Its centralizer in $\mathbf{GL}_2$ consists of all the matrices $\begin{pmatrix} a & b \\ 0 & a \end{pmatrix}$ with $a \in \mathbb{F}_q^\times$ and $b \in \mathbb{F}_q$, so is connected.

But its intersection with $\mathbf{SL}_2$ is not connected if the characteristic is different from 2: it has two connected components corresponding respectively to $a = 1$ and to $a = -1$; so we have $C_{\mathbf{G}}(u)/C_{\mathbf{G}}(u)^\circ \simeq \mathbb{Z}/2\mathbb{Z}$ and the intersection of the geometric conjugacy class of u with $\mathbf{SL}_2^F$ splits into two conjugacy classes. By 3.21 and an easy computation, one sees that the class in $\mathbf{SL}_2^F$ of the matrix $\begin{pmatrix} 1 & a \\ 0 & 1 \end{pmatrix}$ with $a \in \mathbb{F}_q$ depends only on the image of a in $\mathbb{F}_q^\times/(\mathbb{F}_q^\times)^2$.

(iii) An analogous argument shows that, if the characteristic is not 2, the centralizer in $\mathbf{PGL}_2(\overline{\mathbb{F}}_q)$ of the element represented in $\mathbf{GL}_2(\overline{\mathbb{F}}_q)$ by the matrix $\begin{pmatrix} 1 & 0 \\ 0 & -1 \end{pmatrix}$ has two connected components which consist respectively of the matrices of the form $\begin{pmatrix} a & 0 \\ 0 & b \end{pmatrix}$ and of the matrices of the form $\begin{pmatrix} 0 & a \\ b & 0 \end{pmatrix}$; a representative in $\mathbf{GL}_2$ of an element of $\mathbf{PGL}_2^F$ geometrically, but not rationally, conjugate to $\begin{pmatrix} 1 & 0 \\ 0 & -1 \end{pmatrix}$ is $\begin{pmatrix} 0 & \lambda^{-1} \\ \lambda & 0 \end{pmatrix}$ where $\lambda \in \mathbb{F}_{q^2}$ is such that $\lambda^{q-1} = -1$; let us note that any representative in $\mathbf{GL}_2^F$ of this element is not (geometrically) conjugate to $\begin{pmatrix} 1 & 0 \\ 0 & -1 \end{pmatrix}$.

References General results about $\mathbb{F}_q$-structures can be found for example in [P. DELIGNE La conjectures de Weil II, *Publications mathématiques de l'IHES*, **43** (1980), 137–252]. The Lang-Steinberg theorem 3.10 is proved in [St1]; in this paper Steinberg proves the main rationality properties of reductive groups over finite fields. Rationality properties in a more general setting (any field) are studied in [BT].

4. GENERALIZED INDUCTION ASSOCIATED TO A BIMODULE; HARISH-CHANDRA INDUCTION AND RESTRICTION.

A heuristic which is often useful in building the irreducible representations of a finite group G is to find a family $\mathcal{F}$ of subgroups of G "of the same type as G" which generate G, and to build representations of G "by induction" from those of $H \in \mathcal{F}$ (using for instance Ind_H^G; a typical example is the construction of the irreducible representations of the symmetric groups $\mathcal{S}_n$ as $\mathbb{Z}$-linear combinations of $\mathrm{Ind}_{\mathcal{S}_{n_1}\times\ldots\times\mathcal{S}_{n_k}}^{\mathcal{S}_n}$ where $n_1+\ldots+n_k=n$).

In the case of a connected reductive algebraic group defined over $\mathbb{F}_q$, a suitable family $\mathcal{F}$ of subgroups consists of the groups of rational points of rational Levi subgroups of rational parabolic subgroups of $\mathbf{G}$, partially ordered by inclusion.

4.1 PROPOSITION.

(i) *Let* $\mathbf{Q}$ *and* $\mathbf{P}$ *be two parabolic subgroups of* $\mathbf{G}$ *such that* $\mathbf{Q}\subset\mathbf{P}$, *then for any Levi subgroup* $\mathbf{M}$ *of* $\mathbf{Q}$, *there is a unique Levi subgroup* $\mathbf{L}$ *of* $\mathbf{P}$ *such that* $\mathbf{L}\supset\mathbf{M}$.

(ii) *For a Levi subgroup* $\mathbf{L}$ *of a parabolic subgroup* $\mathbf{P}$ *of* $\mathbf{G}$, *the following are equivalent:*

(a) $\mathbf{M}$ *is a Levi subgroup of a parabolic subgroup of* $\mathbf{L}$;

(b) $\mathbf{M}$ *is a Levi subgroup of a parabolic subgroup of* $\mathbf{G}$, *and* $\mathbf{M}\subset\mathbf{L}$.

PROOF: Let us prove (i); given a maximal torus $\mathbf{T}$ of $\mathbf{M}$ there is by 1.17 a unique Levi subgroup $\mathbf{L}$ of $\mathbf{P}$ containing $\mathbf{T}$. It is enough to see that $\mathbf{L}\supset\mathbf{M}$; this is a consequence of 2.1 (i): first notice that with the notation of that proposition we have $\mathbf{U}\subset\mathbf{V}$ (indeed $\mathbf{U}$ is in any Borel subgroup of $\mathbf{P}$, so in any Borel subgroup of $\mathbf{Q}$, hence $R_u(\mathbf{P})\subset\mathbf{Q}$ which implies $R_u(\mathbf{P})\subset R_u(\mathbf{Q})$), thus $(\mathbf{P}\cap\mathbf{Q}).\mathbf{U}=\mathbf{Q}$ and 2.1 (ii) implies the result.

Let us prove (ii); if $\mathbf{M}$ is a Levi subgroup of $\mathbf{Q}_0$, a parabolic subgroup of $\mathbf{L}$, and if $\mathbf{P}=\mathbf{L}\mathbf{U}$, then $\mathbf{Q}_0\mathbf{U}$ is a parabolic subgroup of $\mathbf{G}$ (it is a group since $\mathbf{L}$, thus $\mathbf{Q}_0$, normalizes $\mathbf{U}$, and it clearly contains a Borel subgroup of $\mathbf{G}$); if $\mathbf{V}_0$ is the unipotent radical of $\mathbf{Q}_0$ then clearly $\mathbf{V}_0\mathbf{U}\subset R_u(\mathbf{Q}_0\mathbf{U})$, so $\mathbf{M}$ is a Levi subgroup of $\mathbf{Q}_0\mathbf{U}$ since $\mathbf{Q}_0\mathbf{U}=\mathbf{M}\mathbf{V}_0\mathbf{U}$. Thus (a) implies (b). The converse is an immediate consequence of 2.2, (ii) and (iii). ∎

In the situation (ii) we will say (improperly) that "**M** is a Levi subgroup of **L**".

We will now classify **G**-conjugacy classes of Levi subgroups and $\mathbf{G}^F$-conjugacy classes of rational Levi subgroups; we will really use this classification only in chapter 13.

4.2 PROPOSITION. *Let Φ be the set of roots of* **G** *relative to some maximal torus* **T**, *and let Π be a basis of Φ. The* **G**-*conjugacy classes of Levi subgroups of* **G** *are in one-to-one correspondence with the $W(\mathbf{T})$-orbits of subsets of Π, which correspond themselves one-to-one to the $W(\mathbf{T})$-conjugacy classes of parabolic subgroups of $W(\mathbf{T})$.*

PROOF: Let **B** be the Borel subgroup containing **T** which corresponds to Π. Since all Borel subgroups of **G** are conjugate, and all maximal tori in a parabolic subgroup are conjugate, all Levi subgroups of **G** are conjugate to the Levi subgroup containing **T** of some parabolic subgroup containing **B**, *i.e.*, some $\mathbf{P}_J$. Thus the question becomes one of finding when $\mathbf{L}_J$ is **G**-conjugate to $\mathbf{L}_I$ for two subsets I and J of Π. If $\mathbf{L}_J = {}^g\mathbf{L}_I$ for some $g \in \mathbf{G}$, then, since ${}^{g^{-1}}\mathbf{T}$ and $\mathbf{T}$ are two maximal tori of $\mathbf{L}_I$, there exists $l \in \mathbf{L}_I$ such that ${}^{g^{-1}}\mathbf{T} = {}^l\mathbf{T}$ and $gl \in N_{\mathbf{G}}(\mathbf{T})$ also conjugates $\mathbf{L}_I$ to $\mathbf{L}_J$; so the **G**-conjugacy classes of $\mathbf{L}_I$ are the same as the $W(\mathbf{T})$-conjugacy classes. Since $\mathbf{L}_I$ is generated by **T** and the $\{\mathbf{U}_\alpha\}_{\alpha\in\Phi_I}$, the element $w \in W$ conjugates $\mathbf{L}_I$ to $\mathbf{L}_J$ if and only if ${}^w\Phi_I = \Phi_J$. Since any two bases of Φ_I are conjugate by an element of W_I (by, *e.g.*, 0.31 (v) and 1.5), we may assume that ${}^wI = J$ whence the first part of the statement. To see the second part it is enough to see that if $w \in N_W(W_I)$ then ${}^w\Phi_I = \Phi_I$. This results from the fact that if a reflection s_α is in W_I, then $\alpha \in \Phi_I$: to see that, notice that any element of W_I stabilizes the subspace of $X(\mathbf{T}) \otimes \mathbb{R}$ spanned by I; for the reflection s_α with respect to the vector α to do that, α has to be either in that subspace (in which case $\alpha \in \Phi_I$ by definition), or orthogonal to that subspace. This last case is impossible since then s_α would fix any element of Φ_I which would imply by 0.29 that $l(s_\alpha) = 0$, *i.e.*, $s_\alpha = 1$. ∎

The proof above shows that $N_{\mathbf{G}}(\mathbf{L}_I)/\mathbf{L}_I$ is isomorphic to $N_W(W_I)/W_I$.

4.3 PROPOSITION. *Assume in addition to the hypotheses of 4.2 that* **G** *is defined over $\mathbb{F}_q$ with corresponding Frobenius F and that* **T** *is rational. Then the $\mathbf{G}^F$-classes of rational Levi subgroups of* **G** *are parametrized by F-conjugacy classes of cosets $W_I w$ where $I \subset \Pi$ and w are such that ${}^{wF}W_I = W_I$.*

PROOF: According to 3.21 the $\mathbf{G}^F$-classes of Levi subgroups conjugate under **G** to **L** are parametrized by the F-classes of $N_{\mathbf{G}}(\mathbf{L})/\mathbf{L}$. We may conjugate $\mathbf{L}$ to some $\mathbf{L}_I$ by conjugating a maximal torus of $\mathbf{L}$ to $\mathbf{T}$ but then, as remarked after 3.24, the action of F on $\mathbf{L}$ becomes that of vF on $\mathbf{L}_I$ for some $v \in W(\mathbf{T})$; and the condition that $\mathbf{L}$ is rational becomes ${}^{vF}\mathbf{L}_I = \mathbf{L}_I$, which is equivalent to ${}^{vF}W_I = W_I$ as seen in the proof of 4.2. So the $\mathbf{G}^F$-classes of Levi subgroups conjugate under $\mathbf{G}$ to $\mathbf{L}$ are parametrized by the vF-classes of $N_{\mathbf{G}}(\mathbf{L}_I)/\mathbf{L}_I \simeq N_W(W_I)/W_I$, which is the same as the F-classes under $N_W(W_I)$ of cosets $W_I zv$ where $z \in N_W(W_I)$. This last condition is equivalent to ${}^{zvF}W_I = W_I$. We thus get the proposition by putting $w = zv$; the F-classes under $N_W(W_I)$ are replaced by F-classes under W since, if the F-class of $W_I w$ and that of $W_J w'$ intersect, then L_I is conjugate under $\mathbf{G}$ to L_J. ∎

When $\mathbf{L}$ is a rational Levi subgroup of $\mathbf{G}$, we could, as our introduction suggests, build representations of $\mathbf{G}^F$ from those of $\mathbf{L}^F$ using $\mathrm{Ind}_{\mathbf{L}^F}^{\mathbf{G}^F}$. Actually the representations thus obtained would not have the right properties, in particular their decomposition into irreducible representations would be rather intractable. The right construction is to use the "Harish-Chandra induction" that we are going to describe now as a generalized induction associated to a bimodule.

Let G and H be two finite groups; we will call "G-module-H" a bimodule M endowed with a left $\mathbb{C}[G]$-action and a right $\mathbb{C}[H]$-action (we will view it also sometimes as a $G \times H^{\mathrm{opp}}$-module where H^{opp} denotes the opposed group to H). Such a module gives a functor from the category of left $\mathbb{C}[H]$-modules to that of left $\mathbb{C}[G]$-modules defined by

$$R_H^G : E \mapsto M \otimes_{\mathbb{C}[H]} E$$

where G acts on $M \otimes_{\mathbb{C}[H]} E$ through its action on M. It is clear that the dual module $M^* = \mathrm{Hom}(M, \mathbb{C})$ gives the adjoint functor, which we will denote by ${}^*R_H^G$. The following proposition is immediate from the associativity of the tensor product.

4.4 PROPOSITION (transitivity). *Let G, H, and K be finite groups; let M be a G-module-H and N a H-module-K, then the functor $R_H^G \circ R_K^H$ is equal to the functor R_K^G defined by the G-module-K given by $M \otimes_{\mathbb{C}[H]} N$.*

The modules we will consider in what follows will always be $\mathbb{C}$-vector spaces (or virtual vector spaces) of finite dimension. This hypothesis is necessary in particular for the next proposition which generalizes the character formula for induced characters.

4.5 PROPOSITION. *Let M be a G-module-H and E be an H-module; then for $g \in G$ we have*

$$\operatorname{Trace}(g \mid R_H^G E) = |H|^{-1} \sum_{h \in H} \operatorname{Trace}((g, h^{-1}) \mid M) \operatorname{Trace}(h \mid E).$$

PROOF: The element $|H|^{-1} \sum_h h^{-1} \otimes h$ is an idempotent of the algebra $\mathbb{C}[H \times H^{\mathrm{opp}}] = \mathbb{C}[H] \otimes \mathbb{C}[H^{\mathrm{opp}}]$. Its image in the representation of $H \times H^{\mathrm{opp}}$ on the tensor product $M \otimes_{\mathbb{C}} E$ is a projector whose kernel is generated by the elements of the form $mh \otimes x - m \otimes hx$, since if $\sum_i \sum_h m_i h^{-1} \otimes hx_i = 0$ then

$$\sum_i m_i \otimes x_i = |H|^{-1} \sum_h \sum_i (m_i \otimes x_i - m_i h^{-1} \otimes hx_i).$$

As $M \otimes_{\mathbb{C}[H]} E$ is the quotient of $M \otimes_{\mathbb{C}} E$ by the subspace generated by elements of the form $mh \otimes x - m \otimes hx$, we get

$$\operatorname{Trace}(g \mid R_H^G E) = \operatorname{Trace}(|H|^{-1} \sum_h (g, h^{-1}) \otimes h \mid M \otimes_{\mathbb{C}} E),$$

whence the proposition. ■

4.6 EXAMPLES.

(i) Induction. This is the case where H is a subgroup of G and where M is the group algebra of G on which G acts by left translations and H by right translations.

(ii) Restriction. This is the adjoint functor of induction. In that case M is the group algebra of G on which H acts by left translations and G by right translations.

The next example will play an important part in this book.

(iii) Harish-Chandra induction and restriction. Let $\mathbf{G}$ be a reductive algebraic group defined over $\mathbb{F}_q$; let $\mathbf{P}$ be a rational parabolic subgroup of $\mathbf{G}$ and $\mathbf{L}$ a rational Levi subgroup of $\mathbf{P}$, so that we have a rational Levi decomposition $\mathbf{P} = \mathbf{L}\mathbf{U}$. We take for G the group of rational points $\mathbf{G}^F$, for H the group $\mathbf{L}^F$, and for M the $\mathbf{G}^F$-module-$\mathbf{L}^F$ given by $\mathbb{C}[\mathbf{G}^F/\mathbf{U}^F]$, on which $\mathbf{G}^F$ acts by left translations and $\mathbf{L}^F$ by right translations (which is possible since $\mathbf{L}$ normalizes $\mathbf{U}$). The functor $R_{\mathbf{L}}^{\mathbf{G}}$ thus obtained is called **Harish-Chandra induction**. The adjoint functor, called **Harish-Chandra restriction** and denoted by ${}^*R_{\mathbf{L}}^{\mathbf{G}}$, corresponds to the $\mathbf{L}^F$-module-$\mathbf{G}^F$ given by $\mathbb{C}[\mathbf{U}^F \backslash \mathbf{G}^F]$ where $\mathbf{G}^F$ acts by right translations and $\mathbf{L}^F$ by left translations. Applying 4.5 we get for $\chi \in \mathrm{Irr}(\mathbf{G}^F)$ and $l \in \mathbf{L}^F$

$$\begin{aligned}{}^*R_{\mathbf{L}}^{\mathbf{G}}(\chi)(l) &= |\mathbf{G}^F|^{-1} \sum_{g\in\mathbf{G}^F} \#\{\, x\mathbf{U}^F \in \mathbf{G}^F/\mathbf{U}^F \mid x^{-1}gx \in l\mathbf{U}^F \,\}\chi(g) \\ &= |\mathbf{U}^F|^{-1} \sum_{u\in\mathbf{U}^F} \chi(lu),\end{aligned}$$

where the last equality holds since χ is a central function on $\mathbf{G}^F$ so satisfies $\chi(g) = \chi(x^{-1}gx)$.

NOTATION. We will use the more precise notation $R_{\mathbf{L}^F}^{\mathbf{G}^F}$ and ${}^*R_{\mathbf{L}^F}^{\mathbf{G}^F}$ for Harish-Chandra induction and restriction if there is any possible ambiguity about which Frobenius endomorphism we consider. The parabolic subgroup used in the construction does not appear in the notation since we will prove that $R_{\mathbf{L}}^{\mathbf{G}}$ and ${}^*R_{\mathbf{L}}^{\mathbf{G}}$ do not depend on the parabolic subgroup used in their construction. Until we prove that, to specify which parabolic subgroup has been used, we will use the notation $R_{\mathbf{L}\subset\mathbf{P}}^{\mathbf{G}}$.

NOTE. We have only defined here a functor $R_{\mathbf{L}}^{\mathbf{G}}$ when $\mathbf{L}$ is a rational Levi subgroup of a *rational* parabolic subgroup $\mathbf{P}$ of $\mathbf{G}$; later we will define Deligne-Lusztig induction which generalizes Harish-Chandra induction to the case of non-rational parabolic subgroups $\mathbf{P}$.

The first property to expect of Harish-Chandra induction is the following.

4.7 PROPOSITION (Transitivity of $R_{\mathbf{L}}^{\mathbf{G}}$). *Let $\mathbf{P}$ be a rational parabolic subgroup of $\mathbf{G}$, and $\mathbf{Q}$ a rational parabolic subgroup contained in $\mathbf{P}$. Let $\mathbf{L}$ be a rational Levi subgroup of $\mathbf{P}$ and $\mathbf{M}$ a rational Levi subgroup of $\mathbf{Q}$ contained in $\mathbf{L}$. Then*

$$R_{\mathbf{L}\subset\mathbf{P}}^{\mathbf{G}} \circ R_{\mathbf{M}\subset\mathbf{L}\cap\mathbf{Q}}^{\mathbf{L}} = R_{\mathbf{M}\subset\mathbf{Q}}^{\mathbf{G}}.$$

PROOF: Let $\mathbf{U}$ be the unipotent radical of $\mathbf{P}$ and $\mathbf{V}$ that of $\mathbf{Q}$. We have $\mathbf{U} \subset \mathbf{V}$, so according to 2.1 (iii) we have $\mathbf{V} = \mathbf{U}(\mathbf{L} \cap \mathbf{V})$ and the uniqueness in that decomposition implies that $\mathbf{V}^F = \mathbf{U}^F(\mathbf{L}^F \cap \mathbf{V}^F)$. Using 4.4, proving the proposition is equivalent to showing that

$$\mathbb{C}[\mathbf{G}^F/\mathbf{U}^F] \otimes_{\mathbb{C}[\mathbf{L}^F]} \mathbb{C}[\mathbf{L}^F/(\mathbf{L}\cap\mathbf{V})^F] \xrightarrow{\sim} \mathbb{C}[\mathbf{G}^F/\mathbf{V}^F]$$

as $\mathbf{G}^F$-modules-$\mathbf{M}^F$. But this clearly results from the isomorphism of "$\mathbf{G}^F$-sets-$\mathbf{M}^F$" (with notation of 0.42)

$$\mathbf{G}^F/\mathbf{U}^F \times_{\mathbf{L}^F} \mathbf{L}^F/(\mathbf{L}\cap\mathbf{V})^F \xrightarrow{\sim} \mathbf{G}^F/\mathbf{V}^F$$

which is induced by the map $(g\mathbf{U}^F, l(\mathbf{L}\cap\mathbf{V})^F) \mapsto gl\mathbf{V}^F$ from $\mathbf{G}^F/\mathbf{U}^F \times \mathbf{L}^F/(\mathbf{L}\cap\mathbf{V})^F$ to $\mathbf{G}^F/\mathbf{V}^F$. This map is well-defined since l normalizes $\mathbf{U}$ and

$\mathbf{U} \subset \mathbf{V}$. Using the decomposition $\mathbf{V}^F = \mathbf{U}^F(\mathbf{L} \cap \mathbf{V})^F$, it is easy to see that it factors through the amalgamated product and defines an isomorphism. ■

References

The approach to generalized induction functors using bimodules that we follow here is due to Broué. The construction of 4.6 (iii) was first considered by Harish-Chandra [HC].

5. THE MACKEY FORMULA.

This chapter is devoted to the proof of a fundamental property of Harish-Chandra induction and restriction (5.1 below), which is an analogue of the usual Mackey formula for the composition of a restriction and an induction. As a consequence we will show that Harish-Chandra induction and restriction are independent of the parabolic subgroup used in their definition.

We give this formula below in a form which will remain valid for the Lusztig induction and restriction functors (apart from the rationality of the parabolic subgroups involved) in the cases that we will be able to handle (see chapter 11).

5.1 THEOREM ("MACKEY FORMULA"). *Let* **P** *and* **Q** *be two rational parabolic subgroups of* **G**, *and* **L** *(resp.* **M***) be a rational Levi subgroup of* **P** *(resp.* **Q***). Then*

$$^*R^{\mathbf{G}}_{\mathbf{L}\subset\mathbf{P}} \circ R^{\mathbf{G}}_{\mathbf{M}\subset\mathbf{Q}} = \sum_x R^{\mathbf{L}}_{\mathbf{L}\cap{}^x\mathbf{M}\subset\mathbf{L}\cap{}^x\mathbf{Q}} \circ {}^*R^{{}^x\mathbf{M}}_{\mathbf{L}\cap{}^x\mathbf{M}\subset\mathbf{P}\cap{}^x\mathbf{M}} \circ \operatorname{ad} x,$$

where x runs over a set of representatives of $\mathbf{L}^F\backslash\mathcal{S}(\mathbf{L},\mathbf{M})^F/\mathbf{M}^F$, where

$$\mathcal{S}(\mathbf{L},\mathbf{M}) = \{\, x\in\mathbf{G} \mid \mathbf{L}\cap{}^x\mathbf{M} \text{ contains a maximal torus of } \mathbf{G}\,\}.$$

The notation $\operatorname{ad} x$ above stands for the functor on representations induced by the conjugation by x.

In order to prove this theorem, we first need some information about double coset representatives with respect to parabolic subgroups in **G** and in the Weyl group.

5.2 DEFINITION. *Let W be a Coxeter group with set of generating reflections S, and let $I\subset S$. An element $w\in W$ is called I-***reduced** *(resp.* **reduced**-*I) if for any $v\in W_I$ (the subgroup of W generated by I) we have $l(v)+l(w)=l(vw)$ (resp. $l(w)+l(v)=l(wv)$).*

5.3 LEMMA. *Every coset in $W_I\backslash W$ contains a unique I-reduced element which is the unique element of minimal length in that coset.*

PROOF: Everything else follows easily if we can prove that an element of minimal length in its coset is I-reduced. Suppose otherwise: let w be of minimal length in $W_I w$ and assume there exists $v \in W_I$ such that $l(v) + l(w) < l(vw)$. By adding one by one the terms of a reduced decomposition of v to w and applying the exchange lemma 0.30, we get $vw = \hat{v}\hat{w}$ where $\hat{v}$ (resp. $\hat{w}$) is the product of a proper sub-sequence extracted from a reduced decomposition of v (resp. w). But that would imply that $l(\hat{w}) < l(w)$ which is a contradiction since w and $\hat{w}$ are in the same coset (since $\hat{v} \in W_I$). ■

Of course, the symmetric lemma for reduced-I and W/W_I also holds.

5.4 LEMMA. *Let I and J be two subsets of S.*

(i) *In a double coset in $W_I \backslash W / W_J$, there is a unique element w which is both I-reduced and reduced-J.*

(ii) *The element w of (i) is the unique element of minimal length in the double coset $W_I w W_J$.*

(iii) *Every element of $W_I w W_J$ can be written uniquely as xwy with $l(x) + l(w) + l(y) = l(xwy)$, and where xw is reduced-J.*

An element w such as above is said to be ***I*-reduced-*J***; by symmetry, we may add to the lemma above the statement obtained by replacing in (iii) the condition "xw is reduced-J" by "wy is I-reduced".

PROOF: We first show that two elements w and w' satisfying (i) are of the same length. Let us write $w' = xwy$ with $x \in W_I$ and $y \in W_J$; we get $w'y^{-1} = xw$ and $x^{-1}w' = wy$, whence by definition of w and w' and as $l(y^{-1}) = l(y)$ and $l(x^{-1}) = l(x)$

$$l(w') + l(y) = l(x) + l(w) \text{ and } l(x) + l(w') = l(w) + l(y)$$

whence the result. To get the lemma it is enough now to show that an element of minimal length in its double coset satisfies (iii) (this clearly implies uniqueness). Let $v \in W_I w W_J$ and let xwy be a decomposition with minimal $l(x) + l(y)$ of the (unique) reduced-J element of the coset vW_J. As in the previous lemma, by applying the exchange lemma we find xwy has a reduced decomposition of the form $\hat{x}\hat{w}\hat{y}$ where $\hat{x}$ (resp. $\hat{w}$, $\hat{y}$) is extracted from a reduced decomposition of x (resp. w, y). As before we must have $\hat{w} = w$ otherwise w would not be of minimal length in its double coset. But then the minimality of $l(x) + l(y)$ implies that necessarily $\hat{x} = x$ and $\hat{y} = y$, and so $l(x) + l(w) + l(y) = l(xwy)$; we also get $y = 1$ since otherwise xw would be an element of smaller length than xwy in the coset vW_J. The other properties follow easily. ■

5.5 LEMMA. *Let $\mathbf{G}$ a connected reductive algebraic group, let $\mathbf{T}$ be a maximal torus of $\mathbf{G}$ and let $\mathbf{B}$ be a Borel subgroup containing $\mathbf{T}$. Let $\mathbf{P}_I$ and*

$\mathbf{P}_J$ *be two standard parabolic subgroups (i.e., containing* $\mathbf{B}$*). Then there are canonical bijections*

$$\mathbf{P}_I\backslash\mathbf{G}/\mathbf{P}_J \xrightarrow{\sim} W_I\backslash W/W_J \xrightarrow{\sim} \{\text{Elements of } W \text{ which are } I\text{-reduced-}J\}.$$

PROOF: By the Bruhat decomposition (1.4), any $g \in \mathbf{G}$ is in $\mathbf{B}w\mathbf{B}$ for some $w \in W$, so the double coset $\mathbf{P}_I g \mathbf{P}_J$ is equal to $\mathbf{P}_I w \mathbf{P}_J$. As representatives of W_I (resp. W_J) lie in $\mathbf{P}_I$ (resp. $\mathbf{P}_J$), we may replace w in the above by any element of the double coset $W_I w W_J$. This gives a well-defined surjective map $W_I\backslash W/W_J \to \mathbf{P}_I\backslash\mathbf{G}/\mathbf{P}_J$. This map is injective: assume $\mathbf{P}_I w \mathbf{P}_J = \mathbf{P}_I w' \mathbf{P}_J$; we may assume that w and w' are I-reduced-J. As $\mathbf{P}_I = \mathbf{B}W_I\mathbf{B}$ and $\mathbf{P}_J = \mathbf{B}W_J\mathbf{B}$, we have $\mathbf{B}W_I\mathbf{B}w\mathbf{B}W_J\mathbf{B} = \mathbf{B}W_I\mathbf{B}w'\mathbf{B}W_J\mathbf{B}$. As w and w' are reduced, we have by 1.1 (iii′), using a similar argument to that in the proof of 5.4 that, for any $x \in W_I$ and any $y \in W_J$, $\mathbf{B}x\mathbf{B}w\mathbf{B}y\mathbf{B}$ is a union of sets of the form $\mathbf{B}\hat{x}\mathbf{B}w\mathbf{B}\hat{y}\mathbf{B}$ where $l(\hat{x}w\hat{y}) = l(\hat{x}) + l(w) + l(\hat{y})$; whence $\mathbf{B}W_I\mathbf{B}w\mathbf{B}W_J\mathbf{B} = \mathbf{B}W_I w W_J\mathbf{B}$ and the similar equality for w', whence the equality of the double cosets $W_I w W_J$ and $W_I w' W_J$. The second bijection in the lemma comes from lemma 5.4. ■

We get the following consequence of the above lemma.

5.6 LEMMA. *If* $\mathcal{S}(\mathbf{L},\mathbf{M})$ *is as in 5.1, then:*

(i) The natural map $\mathcal{S}(\mathbf{L},\mathbf{M}) \to \mathbf{P}\backslash\mathbf{G}/\mathbf{Q}$ *induces an isomorphism*

$$\mathbf{L}\backslash\mathcal{S}(\mathbf{L},\mathbf{M})/\mathbf{M} \xrightarrow{\sim} \mathbf{P}\backslash\mathbf{G}/\mathbf{Q}.$$

(ii) $\mathbf{P}^F\backslash\mathbf{G}^F/\mathbf{Q}^F$ *can be identified with the set of rational points of* $\mathbf{P}\backslash\mathbf{G}/\mathbf{Q}$*, and* $\mathbf{L}^F\backslash\mathcal{S}(\mathbf{L},\mathbf{M})^F/\mathbf{M}^F$ *with the set of rational points of* $\mathbf{L}\backslash\mathcal{S}(\mathbf{L},\mathbf{M})/\mathbf{M}$.

PROOF: Let us first prove (ii): by 3.21, since $\mathbf{P}\times\mathbf{Q}$, as well as the stabilizer $\mathbf{P}\cap{}^x\mathbf{Q}$ of a point $x \in \mathbf{G}$ under the action of $\mathbf{P}\times\mathbf{Q}$, is connected, the double cosets $\mathbf{P}^F\backslash\mathbf{G}^F/\mathbf{Q}^F$ can be identified with the rational double cosets in $\mathbf{P}\backslash\mathbf{G}/\mathbf{Q}$. Similarly, as $\mathbf{L}\cap{}^x\mathbf{M}$ is connected when it contains a maximal torus (see 2.1), the double cosets $\mathbf{L}^F\backslash\mathcal{S}(\mathbf{L},\mathbf{M})^F/\mathbf{M}^F$ can be identified with the rational double cosets in $\mathbf{L}\backslash\mathcal{S}(\mathbf{L},\mathbf{M})/\mathbf{M}$.

Let us prove (i). We may conjugate $\mathbf{Q}$ by an element $g \in \mathbf{G}$ so that $\mathbf{P}$ and ${}^g\mathbf{Q}$ have a common Borel subgroup and that $\mathbf{L}$ and ${}^g\mathbf{M}$ have a common maximal torus $\mathbf{T}$. Then for any $x \in \mathbf{G}$ we have $\mathbf{P}x\mathbf{Q} = (\mathbf{P}xg^{-1}{}^g\mathbf{Q})g$, *i.e.*, the double cosets with respect to $\mathbf{P}$ and $\mathbf{Q}$ are translates of those with respect to $\mathbf{P}$ and ${}^g\mathbf{Q}$. Similarly $\mathcal{S}(\mathbf{L},\mathbf{M}) = \mathcal{S}(\mathbf{L},{}^g\mathbf{M})g$. Thus to show (i) we may replace $\mathbf{Q}$ by ${}^g\mathbf{Q}$, *i.e.*, assume that $\mathbf{P}$ and $\mathbf{Q}$ are standard and that $\mathbf{L}$ and $\mathbf{M}$ are standard Levi subgroups, *i.e.*, $\mathbf{P} = \mathbf{P}_I$, $\mathbf{Q} = \mathbf{P}_J$, $\mathbf{L} = \mathbf{L}_I$, and $\mathbf{M} = \mathbf{L}_J$. By lemma 5.5, there exist representatives in $N(\mathbf{T})$ of the double cosets in $\mathbf{P}_I\backslash\mathbf{G}/\mathbf{P}_J$. Thus every double coset has a representative in $\mathcal{S}(\mathbf{L},\mathbf{M})$ and the map of (i) is surjective. Conversely, if $x \in \mathcal{S}(\mathbf{L},\mathbf{M})$, the group $\mathbf{L}\cap{}^x\mathbf{M}$ contains a maximal torus; this torus is of the form ${}^l\mathbf{T}$ for some $l \in \mathbf{L}$ and also of the form ${}^{xm}\mathbf{T}$, for some $m \in \mathbf{M}$, since all maximal tori in an algebraic group are conjugate. But then $l^{-1}xm \in N(\mathbf{T})$ and is in the same double coset as x with respect to $\mathbf{L}$ and $\mathbf{M}$. Let now x and y be two elements of $\mathcal{S}(\mathbf{L},\mathbf{M})$ such that $\mathbf{P}x\mathbf{Q} = \mathbf{P}y\mathbf{Q}$. As we have just seen, we may translate x and y by elements of $\mathbf{L}$ and $\mathbf{M}$ so they fall into $N_{\mathbf{G}}(\mathbf{T})$. By lemma 5.5, the images of x and y are in the same double coset with respect to W_I and W_J, so x and y represent the same double coset with respect to $\mathbf{L}$ and $\mathbf{M}$. ∎

PROOF OF THEOREM 5.1: Let $\mathbf{U}$ (resp. $\mathbf{V}$) denote the unipotent radical of $\mathbf{P}$ (resp. $\mathbf{Q}$). By 4.4, the functor in the left-hand side of the Mackey formula corresponds to the $\mathbf{L}$-module-$\mathbf{M}$ given by

$$\mathbb{C}[\mathbf{U}^F\backslash\mathbf{G}^F]\otimes_{\mathbb{C}[\mathbf{G}^F]}\mathbb{C}[\mathbf{G}^F/\mathbf{V}^F].$$

This module is clearly isomorphic to $\mathbb{C}[\mathbf{U}^F\backslash\mathbf{G}^F/\mathbf{V}^F]$. Let us decompose $\mathbf{G}^F$ into double cosets with respect to $\mathbf{P}^F$ and $\mathbf{Q}^F$. We get

$$\mathbf{U}^F\backslash\mathbf{G}^F/\mathbf{V}^F = \coprod_{x\in\mathbf{L}^F\backslash\mathcal{S}(\mathbf{L},\mathbf{M})^F/\mathbf{M}^F}\mathbf{U}^F\backslash\mathbf{P}^F x\mathbf{Q}^F/\mathbf{V}^F.$$

We now use

5.7 LEMMA. *For any $x \in \mathcal{S}(\mathbf{L},\mathbf{M})^F$ the map*

$$(l(\mathbf{L}\cap{}^x\mathbf{V})^F, ({}^x\mathbf{M}\cap\mathbf{U})^F.{}^xm) \mapsto \mathbf{U}^F lxm\mathbf{V}^F$$

is an isomorphism

$$\mathbf{L}^F/(\mathbf{L}\cap{}^x\mathbf{V})^F \times_{(\mathbf{L}\cap{}^x\mathbf{M})^F} ({}^x\mathbf{M}\cap\mathbf{U})^F\backslash{}^x\mathbf{M}^F \xrightarrow{\sim} \mathbf{U}^F\backslash\mathbf{P}^F x\mathbf{Q}^F/\mathbf{V}^F,$$

where the notation is as in 0.42.

PROOF: It is easy to check that the map is well-defined. As the stabilizer $\mathbf{U}\cap{}^x\mathbf{V}$ of a point $x \in \mathbf{G}$ under the action of $\mathbf{U}\times\mathbf{V}$ is always connected (see 0.33), and as the stabilizer of a point of $\mathbf{L}^F/(\mathbf{L}\cap{}^x\mathbf{V})^F \times ({}^x\mathbf{M}\cap\mathbf{U})^F\backslash{}^x\mathbf{M}^F$ under the diagonal action of $\mathbf{L}\cap{}^x\mathbf{M}$ is reduced to 1, and thus connected, it is enough to show that the same map at the level of the underlying algebraic varieties is an isomorphism, as it clearly commutes with F. The map is surjective since any element of $\mathbf{P}x\mathbf{Q}$ has a decomposition of the form $ulxmv$ (with each term in the group suggested by its name). We now prove its injectivity. If $\mathbf{U}lxm\mathbf{V} = \mathbf{U}l'xm'\mathbf{V}$, then $lxm = l'uxvm'$ for some $u \in \mathbf{U}$ and $v \in \mathbf{V}$ since l' normalizes $\mathbf{U}$ and m' normalizes $\mathbf{V}$. Thus

$$u^{-1}l'^{-1}l = {}^x(vm'm^{-1}) \in \mathbf{P} \cap {}^x\mathbf{Q}. \tag{1}$$

The component in $\mathbf{L} \cap {}^x\mathbf{M}$ of the left-hand side of (1) is that of $l'^{-1}l$, and the component in $\mathbf{L} \cap {}^x\mathbf{M}$ of the right-hand side of (1) is that of ${}^x(m'm^{-1})$ by the following

5.8 LEMMA. *Let $\mathbf{L}$ and $\mathbf{M}$ be respective Levi subgroups of two parabolic subgroups $\mathbf{P}$ and $\mathbf{Q}$, and assume they have a common maximal torus. If $ul = vm \in \mathbf{P} \cap \mathbf{Q}$, then $u_{\mathbf{M}} = m_{\mathbf{U}}$, $l_{\mathbf{V}} = v_{\mathbf{L}}$ and $l_{\mathbf{M}} = m_{\mathbf{L}}$, with obvious notation: $u_{\mathbf{M}}$ is the component in $\mathbf{U} \cap \mathbf{M}$ of $u \in \mathbf{U}$ in the decomposition $(\mathbf{U} \cap \mathbf{V})(\mathbf{U} \cap \mathbf{M})(\mathbf{L} \cap \mathbf{V})(\mathbf{L} \cap \mathbf{M})$ of 2.1 (iii), etc.*

PROOF: We have

$$u_{\mathbf{V}}u_{\mathbf{M}}l_{\mathbf{V}}l_{\mathbf{M}} = v_{\mathbf{U}}v_{\mathbf{L}}m_{\mathbf{U}}m_{\mathbf{L}} = v_{\mathbf{U}}[v_{\mathbf{L}}, m_{\mathbf{U}}]m_{\mathbf{U}}v_{\mathbf{L}}m_{\mathbf{L}}$$

and the commutator which appears in this formula is in $\mathbf{V} \cap \mathbf{U}$. The uniqueness of the decomposition in 2.1 (iii) gives the result. ■

Thus $(l(\mathbf{L} \cap {}^x\mathbf{V}), ({}^x\mathbf{M} \cap \mathbf{U})^x m)$ and $(l'(\mathbf{L} \cap {}^x\mathbf{V}), ({}^x\mathbf{M} \cap \mathbf{U})^x m')$ are equal in the amalgamated product, whence the injectivity and lemma 5.7. ■

Taking then the union over all x, we get

$$\mathbf{U}^F \backslash \mathbf{G}^F / \mathbf{V}^F \xrightarrow{\sim} \coprod_{x \in \mathbf{L}^F \backslash \mathcal{S}(\mathbf{L},\mathbf{M})^F / \mathbf{M}^F} \mathbf{L}^F / (\mathbf{L} \cap {}^x\mathbf{V})^F \times_{(\mathbf{L} \cap {}^x\mathbf{M})^F} ({}^x\mathbf{M} \cap \mathbf{U})^F \backslash {}^x\mathbf{M}^F,$$

and this bijection is compatible with the action of $\mathbf{L}^F$ by left multiplication, and with that of $\mathbf{M}^F$ by right multiplication on the left-hand side and by the composite of left multiplication and ad x on the x term of the right-hand side. We finally get an isomorphism of $\mathbf{L}^F$-modules-$\mathbf{M}^F$:

$$\mathbb{C}[\mathbf{U}^F \backslash \mathbf{G}^F / \mathbf{V}^F] \xrightarrow{\sim} \coprod_{x \in \mathbf{L}^F \backslash \mathcal{S}(\mathbf{L},\mathbf{M})^F / \mathbf{M}^F} \mathbb{C}[\mathbf{L}^F / (\mathbf{L} \cap {}^x\mathbf{V})^F] \otimes_{\mathbb{C}[(\mathbf{L} \cap {}^x\mathbf{M})^F]} \mathbb{C}[({}^x\mathbf{M} \cap \mathbf{U})^F \backslash {}^x\mathbf{M}^F],$$

where the action of $\mathbf{M}^F$ on the right-hand side is as explained above. The $\mathbf{L}^F$-module-$\mathbf{M}^F$ which appears in the right-hand side is exactly the one which corresponds to the functor in the right-hand side of the Mackey formula, whence the theorem. ■

References

The notion of I-reduced elements in a Coxeter group can be found with additional information in the exercises of [Bbk, IV §1]. The Mackey formula 5.1 was first stated by Deligne but first published in [G. Lusztig and N. Spaltenstein Induced unipotent classes, *J. of London Mathematical Society,* **19** (1979), 41–52, 2.5].

6. HARISH-CHANDRA'S THEORY

We explain here the theory of "cuspidal representations" which is due to Harish-Chandra. First, as announced, we prove:

6.1 PROPOSITION. *The functor* $R^{\mathbf{G}}_{\mathbf{L}\subset\mathbf{P}}$ *is independent of* $\mathbf{P}$.

Thus we are allowed to omit $\mathbf{P}$ from the notation.

PROOF: The proof is by induction on the **semi-simple rank** of $\mathbf{G}$, which is by definition the rank of $\mathbf{G}/R\mathbf{G}$.

Let us note that the semi-simple rank of a Levi component of a proper parabolic subgroup of $\mathbf{G}$ is strictly less than that of $\mathbf{G}$ by 0.17, 0.37 (iii) and 1.15. If $\mathbf{L} = \mathbf{G}$ there is nothing to prove, so we may assume that the semi-simple rank of $\mathbf{L}$ is strictly less than that of $\mathbf{G}$. By the Mackey formula applied to only one Levi subgroup $\mathbf{L}$ but with two different parabolic subgroups $\mathbf{P}$ and $\mathbf{Q}$, we get

$$\langle R^{\mathbf{G}}_{\mathbf{L}\subset\mathbf{P}}\pi, R^{\mathbf{G}}_{\mathbf{L}\subset\mathbf{Q}}\pi \rangle_{\mathbf{G}^F} = \sum_{x\in\mathbf{L}^F\backslash\mathcal{S}(\mathbf{L},\mathbf{L})^F/\mathbf{L}^F} \langle {}^*R^{{}^x\mathbf{L}}_{\mathbf{L}\cap{}^x\mathbf{L}\subset\mathbf{P}\cap{}^x\mathbf{L}}{}^x\pi, {}^*R^{\mathbf{L}}_{\mathbf{L}\cap{}^x\mathbf{L}\subset\mathbf{L}\cap{}^x\mathbf{Q}}\pi \rangle_{\mathbf{L}^F\cap{}^x\mathbf{L}^F}.$$

By induction, the right-hand side does not depend on the parabolic subgroups $\mathbf{P}$ and $\mathbf{Q}$, so the same is true for the left-hand side which means that, if we put $f(\mathbf{P}) = R^{\mathbf{G}}_{\mathbf{L}\subset\mathbf{P}}\pi$, we have

$$\langle f(\mathbf{P}), f(\mathbf{P}) \rangle_{\mathbf{G}^F} = \langle f(\mathbf{P}), f(\mathbf{Q}) \rangle_{\mathbf{G}^F} = \langle f(\mathbf{Q}), f(\mathbf{Q}) \rangle_{\mathbf{G}^F} = \langle f(\mathbf{Q}), f(\mathbf{P}) \rangle_{\mathbf{G}^F} \quad (6.1.1)$$

whence $\langle f(\mathbf{P}) - f(\mathbf{Q}), f(\mathbf{P}) - f(\mathbf{Q}) \rangle_{\mathbf{G}^F} = 0$ and so $f(\mathbf{P}) = f(\mathbf{Q})$. ■

6.2 EXAMPLE. Let $\mathcal{R}\mathbb{GL} = \bigoplus_{n>=0} \mathcal{R}[\mathbb{GL}_n(\mathbb{F}_q)]$ (where for any finite group H, we denote by $\mathcal{R}(H)$ the Grothendieck group of characters of H). One can define on $\mathcal{R}\mathbb{GL}$ a product by putting, for two irreducible characters $\chi \in \mathrm{Irr}(\mathbb{GL}_n(\mathbb{F}_q))$ and $\psi \in \mathrm{Irr}(\mathbb{GL}_m(\mathbb{F}_q))$,

$$\chi \circ \psi = R^{\mathbb{GL}_{n+m}}_{\mathbb{GL}_n\times\mathbb{GL}_m\subset\mathbf{P}_{n,m}}\chi \otimes \psi,$$

where $\mathbb{GL}_n \times \mathbb{GL}_m$ is embedded in $\mathbb{GL}_{n+m}$ as a group of block-diagonal matrices

$$\begin{pmatrix} \mathbb{GL}_n & 0 \\ 0 & \mathbb{GL}_m \end{pmatrix},$$

and where the parabolic subgroup $\mathbf{P}_{n,m}$ is the group of block-triangular matrices (see chapter 15 below)

$$\begin{pmatrix} \mathbf{GL}_n & * \\ 0 & \mathbf{GL}_m \end{pmatrix}.$$

We thus get an algebra structure and by 6.1 the product is commutative. Let us note that the Levi subgroups $\mathbf{GL}_n \times \mathbf{GL}_m$ and $\mathbf{GL}_m \times \mathbf{GL}_n$ are conjugate in $\mathbf{GL}_{n+m}$ but not so the parabolic subgroups $\mathbf{P}_{n,m}$ and $\mathbf{P}_{m,n}$. Because of this fact, it is difficult to give a direct proof of the commutativity. One can also define on $\mathcal{R}\mathbf{GL}$ a cocommutative coproduct $\mathcal{R}\mathbf{GL} \to \mathcal{R}\mathbf{GL} \otimes \mathcal{R}\mathbf{GL} : \chi \in \operatorname{Irr}(\mathbf{GL}_n(\mathbb{F}_q)) \mapsto \bigoplus_{i+j=n} {}^*R^{\mathbf{GL}_n}_{\mathbf{GL}_i \times \mathbf{GL}_j} \chi$, and it can be shown using the Mackey formula that these two laws define on $\mathcal{R}\mathbf{GL}$ a Hopf algebra structure.

Because of the transitivity of $R^{\mathbf{G}}_{\mathbf{L}}$ (see 4.7), one can define a partial order on the set of pairs $(\mathbf{L}, \delta)$ consisting of a rational Levi subgroup of a rational parabolic subgroup of $\mathbf{G}$, and of $\delta \in \operatorname{Irr}(\mathbf{L}^F)$, by putting $(\mathbf{L}', \delta') \leq (\mathbf{L}, \delta)$ if $\mathbf{L}' \subset \mathbf{L}$ and $\langle\, \delta, R^{\mathbf{L}}_{\mathbf{L}'} \delta' \,\rangle_{\mathbf{L}^F} \neq 0$.

6.3 PROPOSITION. *The following properties are equivalent:*
(i) The pair $(\mathbf{L}, \delta)$ is minimal for the partial order defined above,
(ii) For any rational Levi subgroup $\mathbf{L}'$ of a rational parabolic subgroup of $\mathbf{L}$, we have ${}^*R^{\mathbf{L}}_{\mathbf{L}'} \delta = 0$.

If these conditions are fulfilled, the representation δ of $\mathbf{L}^F$ is said to be **cuspidal**.

PROOF: The pair $(\mathbf{L}, \delta)$ is minimal if and only if for any $\mathbf{L}' \subset \mathbf{L}$ and any $\delta' \in \operatorname{Irr}(\mathbf{L}'^F)$ we have

$$\langle\, \delta, R^{\mathbf{L}}_{\mathbf{L}'} \delta' \,\rangle_{\mathbf{L}^F} = \langle\, {}^*R^{\mathbf{L}}_{\mathbf{L}'} \delta, \delta' \,\rangle_{\mathbf{L}'^F} = 0,$$

which is equivalent to (ii), whence the result. ∎

6.4 THEOREM. *Let $\chi \in \operatorname{Irr}(\mathbf{G}^F)$; then, up to $\mathbf{G}^F$-conjugacy, there exists a unique minimal pair $(\mathbf{L}, \delta)$ such that $(\mathbf{L}, \delta) \leq (\mathbf{G}, \chi)$.*

PROOF: We need the following lemma.

6.5 LEMMA. *If $\delta \in \operatorname{Irr}(\mathbf{L}^F)$ and $\eta \in \operatorname{Irr}(\mathbf{M}^F)$ are two cuspidal representations of rational Levi subgroups of rational parabolic subgroups of $\mathbf{G}$, then*

$$\langle\, R^{\mathbf{G}}_{\mathbf{L}} \delta, R^{\mathbf{G}}_{\mathbf{M}} \eta \,\rangle_{\mathbf{G}^F} = |\{\, x \in \mathbf{G}^F \mid {}^x\mathbf{L} = \mathbf{M} \text{ and } {}^x\delta = \eta \,\}/\mathbf{L}^F|.$$

PROOF: By the Mackey formula, we have

$$\begin{aligned}\langle R_{\mathbf{L}}^{\mathbf{G}}\delta, R_{\mathbf{M}}^{\mathbf{G}}\eta\rangle_{\mathbf{G}^F} &= \langle \delta, {}^*R_{\mathbf{L}}^{\mathbf{G}} R_{\mathbf{M}}^{\mathbf{G}}\eta\rangle_{\mathbf{L}^F}\\ &= \langle \delta, \sum_{x\in\mathbf{L}^F\backslash\mathcal{S}(\mathbf{L},\mathbf{M})^F/\mathbf{M}^F} \operatorname{ad} x^{-1} R_{{}^x\mathbf{L}\cap\mathbf{M}}^{{}^x\mathbf{L}}\, {}^*R_{{}^x\mathbf{L}\cap\mathbf{M}}^{\mathbf{M}}\eta\rangle_{\mathbf{L}^F}\\ &= \langle \delta, \sum_{\{x\in\mathbf{L}^F\backslash\mathcal{S}(\mathbf{L},\mathbf{M})^F/\mathbf{M}^F\mid{}^x\mathbf{L}\supset\mathbf{M}\}} \operatorname{ad} x^{-1} R_{\mathbf{M}}^{{}^x\mathbf{L}}\eta\rangle_{\mathbf{L}^F},\end{aligned}$$

the last equality because, as η is cuspidal, we have ${}^*R_{{}^x\mathbf{L}\cap\mathbf{M}}^{\mathbf{M}}\eta = 0$ if ${}^x\mathbf{L}\cap\mathbf{M}\neq\mathbf{M}$. In the same way we get

$$\langle \delta, \sum_{\{x\in\mathbf{L}^F\backslash\mathcal{S}(\mathbf{L},\mathbf{M})^F/\mathbf{M}^F\mid{}^x\mathbf{L}\supset\mathbf{M}\}} \operatorname{ad} x^{-1} R_{\mathbf{M}}^{{}^x\mathbf{L}}\eta\rangle_{\mathbf{L}^F} =$$

$$\langle \sum_{\{x\in\mathbf{L}^F\backslash\mathcal{S}(\mathbf{L},\mathbf{M})^F/\mathbf{M}^F\mid{}^x\mathbf{L}\supset\mathbf{M}\}} {}^*R_{\mathbf{M}}^{{}^x\mathbf{L}}\,{}^x\delta, \eta\rangle_{\mathbf{M}^F} = \langle \sum_{\{x\in\mathbf{L}^F\backslash\mathcal{S}(\mathbf{L},\mathbf{M})^F/\mathbf{M}^F\mid{}^x\mathbf{L}=\mathbf{M}\}} {}^x\delta, \eta\rangle_{\mathbf{M}^F},$$

the last equality coming from the fact that δ (whence also ${}^x\delta$) is cuspidal, whence the lemma. ■

Theorem 6.4 is a straightforward consequence of lemma 6.5 as, if $\langle \chi, R_{\mathbf{L}}^{\mathbf{G}}\delta\rangle_{\mathbf{G}^F}$ and $\langle \chi, R_{\mathbf{M}}^{\mathbf{G}}\eta\rangle_{\mathbf{G}^F}$ are both different from zero, then $\langle R_{\mathbf{L}}^{\mathbf{G}}\delta, R_{\mathbf{M}}^{\mathbf{G}}\eta\rangle_{\mathbf{G}^F}$ cannot be zero. ■

For $\delta\in\operatorname{Irr}(\mathbf{L}^F)$ we put

$$W_{\mathbf{G}}(\delta) = \{\, w\in N_{\mathbf{G}^F}(\mathbf{L})/\mathbf{L}^F \mid {}^w\delta=\delta\,\}.$$

If δ is cuspidal, by 6.5 we have $\langle R_{\mathbf{L}}^{\mathbf{G}}\delta, R_{\mathbf{L}}^{\mathbf{G}}\delta\rangle_{\mathbf{G}^F} = |W_{\mathbf{G}}(\delta)|$. The following more precise result is true.

6.6 REMARK. A theorem proved by Howlett and Lehrer [Induced cuspidal representations and generalized Hecke rings, *Inventiones Math.*, **58** (1980), 37–64] states that $\operatorname{End}_{\mathbf{G}^F}(R_{\mathbf{L}}^{\mathbf{G}}(\delta)) \simeq \mathbb{C}[W_{\mathbf{G}}(\delta)]^{\mu}$, where the exponent μ means "twisted by a cocycle μ" (it has been proved by Lusztig [L4, 8.6] that this cocycle is trivial when the centre of $\mathbf{G}$ is connected). So the irreducible components of $R_{\mathbf{L}}^{\mathbf{G}}\delta$ are parametrized by $\operatorname{Irr}(W_{\mathbf{G}}(\delta))$, and, if ρ_χ is the component corresponding to $\chi\in\operatorname{Irr}(W_{\mathbf{G}}(\delta))$, we have $\langle \rho_\chi, R_{\mathbf{L}}^{\mathbf{G}}\delta\rangle_{\mathbf{G}^F} = \dim\chi$.

The results in this chapter give a first approach, due to Harish-Chandra, towards the classification of irreducible characters of $\mathbf{G}^F$. From 6.4 above, we get a partition of $\operatorname{Irr}(\mathbf{G}^F)$ in series parametrized by cuspidal representations of rational Levi subgroups of rational parabolic subgroups (up to $\mathbf{G}^F$-conjugacy): the set of irreducible components of $R_{\mathbf{L}}^{\mathbf{G}}\delta$ is called the **Harish-Chandra series** associated to $(\mathbf{L},\delta)$. Sometimes the union of such series when $\mathbf{L}$ is fixed and δ runs over the set of cuspidal representations of $\mathbf{L}^F$ is called the Harish-Chandra series associated to $\mathbf{L}$. When $\mathbf{L}$ is a maximal torus then all irreducible (*i.e.*, degree one) representations of $\mathbf{L}^F$ are cuspidal; all such tori are $\mathbf{G}^F$-conjugate and the series associated to a (any) rational maximal torus (included in a rational Borel subgroup) is called the **principal series**. The set of cuspidal representations of $\mathbf{G}^F$ is also the series associated to $\mathbf{G}$, which is called the **discrete series**. So the first problem is to study the discrete series (not only in $\mathbf{G}$ but in any Levi subgroup of a rational parabolic subgroup of $\mathbf{G}$).

References

Two possible references for Harish-Chandra's theory are the initial paper by Harish-Chandra [HC] and Springer's paper "Cusp forms for finite groups" in [B2]. Zhelevinski's book [Z] explains the theory of representations of linear groups from the viewpoint mentioned above in 6.2.

7. FURTHER RESULTS ON HARISH-CHANDRA INDUCTION

As in the previous chapter we consider a Levi decomposition $\mathbf{P} = \mathbf{L}\mathbf{U}$ of a rational parabolic subgroup of $\mathbf{G}$. Let us begin with a finite group theoretic result.

7.1 PROPOSITION. *Let s be the semi-simple part of an element $l \in \mathbf{L}^F$; then the map $\mathbf{U}^F \times (C_{\mathbf{G}}(s) \cap \mathbf{U}^F) \to l\mathbf{U}^F$ defined by $(y, z) \mapsto {}^y(lz)$ is onto and all its fibres have the same cardinality equal to $|C_{\mathbf{G}}(s) \cap \mathbf{U}^F|$.*

Before proving that result we give some corollaries which show its usefulness.

7.2 NOTATION. *We shall denote by $\mathcal{C}(H)$ the set of class functions on a finite group H and by $\mathcal{C}(H)_{p'}$ the set of class functions f on H such that $f(x) = f(x_{p'})$ for any $x \in H$, where $x_{p'}$ is the p'-part of the element x.*

7.3 COROLLARY. *If $f \in \mathcal{C}(\mathbf{G}^F)_{p'}$, it satisfies $f(lu) = f(l)$ for any $(u, l) \in \mathbf{U}^F \times \mathbf{L}^F$.*

PROOF: By 7.1, if $l = sv$ is the Jordan decomposition of l, there exists $y \in \mathbf{U}^F$ and $z \in C_{\mathbf{G}}(s) \cap \mathbf{U}^F$ such that ${}^y(lz) = lu$, so, in particular $f(lz) = f(lu)$. But, as v normalizes $C_{\mathbf{G}}(s) \cap \mathbf{U}^F$, the element vz is unipotent in $C_{\mathbf{G}}(s)$ and $s.vz$ is the Jordan decomposition of lz; so $f(lz) = f(s) = f(l)$, whence the result. ■

7.4 COROLLARY. *If $f \in \mathcal{C}(\mathbf{G}^F)_{p'}$, then ${}^*R_{\mathbf{L}}^{\mathbf{G}} f = \operatorname{Res}_{\mathbf{L}^F}^{\mathbf{G}^F} f$ and ${}^*R_{\mathbf{L}}^{\mathbf{G}}(\chi f) = ({}^*R_{\mathbf{L}}^{\mathbf{G}}\chi).\operatorname{Res}_{\mathbf{L}^F}^{\mathbf{G}^F} f$ for any $\chi \in \mathcal{C}(\mathbf{G}^F)$.*

PROOF: This result is a straightforward consequence of 7.3 and the formula

$$({}^*R_{\mathbf{L}}^{\mathbf{G}} f)(l) = |\mathbf{U}^F|^{-1} \sum_{u \in \mathbf{U}^F} f(lu).$$

■

7.5 COROLLARY. *If $f \in \mathcal{C}(\mathbf{G}^F)_{p'}$, then $R_{\mathbf{L}}^{\mathbf{G}}(\chi \operatorname{Res}_{\mathbf{L}^F}^{\mathbf{G}^F} f) = (R_{\mathbf{L}}^{\mathbf{G}}\chi).f$ for any $\chi \in \mathcal{C}(\mathbf{G}^F)$.*

This is what we get by taking the adjoint from 7.4. ■

7.6 COROLLARY. *With the above notation, let s be the semi-simple part of an element $l \in \mathbf{L}$; then* $(\mathrm{Res}^{\mathbf{L}^F}_{C^\circ_{\mathbf{L}}(s)^F} {}^*R^{\mathbf{G}}_{\mathbf{L}}\chi)(l) = ({}^*R^{C^\circ_{\mathbf{G}}(s)}_{C^\circ_{\mathbf{L}}(s)} \mathrm{Res}^{\mathbf{G}^F}_{C^\circ_{\mathbf{G}}(s)^F} \chi)(l)$.

PROOF: First let us note that the above formula has a well-defined meaning because $l \in C^\circ_{\mathbf{L}}(s)$ by 2.5, and $C^\circ_{\mathbf{L}}(s)$ is a Levi subgroup of $C^\circ_{\mathbf{G}}(s)$. Indeed $\mathbf{P} \cap C^\circ_{\mathbf{G}}(s)$ is a parabolic subgroup of $C^\circ_{\mathbf{G}}(s)$ with Levi decomposition $(\mathbf{L} \cap C^\circ_{\mathbf{G}}(s))(\mathbf{U} \cap C^\circ_{\mathbf{G}}(s))$ since $C^\circ_{\mathbf{G}}(s)$ is a connected reductive subgroup of maximal rank of $\mathbf{G}$, see 2.3 (ii). So $\mathbf{L} \cap C^\circ_{\mathbf{G}}(s)$ is connected and $\mathbf{L} \cap C^\circ_{\mathbf{G}}(s) = C^\circ_{\mathbf{L}}(s)$.

We have

$$({}^*R^{\mathbf{G}}_{\mathbf{L}}\chi)(l) = |\mathbf{U}^F|^{-1} \sum_{u \in \mathbf{U}^F} \chi(lu) = |C_{\mathbf{G}}(s) \cap \mathbf{U}^F|^{-1} \sum_{z \in C_{\mathbf{G}}(s) \cap \mathbf{U}^F} \chi(lz),$$

the latter equality by 7.1. We then get the result using the fact that $C_{\mathbf{G}}(s) \cap \mathbf{U} = C^\circ_{\mathbf{G}}(s) \cap \mathbf{U}$ (see 2.5 again). ■

PROOF OF 7.1: Let v be the unipotent part of l and let $V = <v, \mathbf{U}^F>$; the group V is the unique p-Sylow subgroup of $<l, \mathbf{U}^F>$ because it is normal and the quotient is cyclic generated by a p'-element (the image of s). For any $u \in \mathbf{U}^F$ we must count the number of solutions in $(y, z) \in \mathbf{U}^F \times (C_{\mathbf{G}}(s) \cap \mathbf{U}^F)$ to the equation ${}^y(lz) = lu$. Let $s'u'$ be the Jordan decomposition of lu; we have seen in the proof of 7.3 that $s.vz$ is the Jordan decomposition of lz, so the equation above is equivalent to the system of two equations ${}^y s = s'$ and ${}^y(vz) = u'$.

But $<s>$ and $<s'>$ are two p'-complements of the normal subgroup V of $<l, \mathbf{U}^F>$ so, by the Schur-Zassenhaus theorem (see [Go, 6.2.1]), they are conjugate by an element of V. As s cannot be conjugate to one of its powers other than itself in the quotient group $<l, \mathbf{U}^F>/V$, there exists $y \in V$ such that ${}^y s = s'$. The element y can be taken in $\mathbf{U}^F$ as v centralizes s. If we prove that the element z defined by $z = v^{-1}.{}^{y^{-1}}u'$ is in $C_{\mathbf{G}}(s) \cap \mathbf{U}^F$ then we are done, as there are $|C_{\mathbf{G}}(s) \cap \mathbf{U}^F|$ solutions y to ${}^y s = s'$ and the value of z is uniquely defined by that of y.

But v and s commute, and ${}^{y^{-1}}u'$ commutes with ${}^{y^{-1}}s' = s$, so z commutes with s. It remains for us to show that z is in $\mathbf{U}^F$: we have $lz = {}^{y^{-1}}(lu) \in {}^{y^{-1}}(l\mathbf{U}^F) = l\mathbf{U}^F$ (this last equality because $y \in \mathbf{U}^F$), whence the result. ■

References

Property 7.6 has been proved and used, under somewhat stronger assumptions, by Curtis in [Cu]; we shall generalize it in chapter 12.

8. THE DUALITY FUNCTOR

This chapter is devoted to the exposition of the main properties of the "duality" functor for the characters of $\mathbf{G}^F$. Though it has been used only since 1981, this functor has an elementary definition and allows substantial simplification in some areas of the representation theory of $\mathbf{G}^F$.

First we need the notion of the $\mathbb{F}_q$-rank of an algebraic group. Recall (see remarks before 3.17) that if $\mathbf{T}$ is a torus defined over $\mathbb{F}_q$ there is an automorphism τ of $X(\mathbf{T})$ such that for any $\alpha \in X(\mathbf{T})$ and $t \in \mathbf{T}$ we have

$$(\tau\alpha)({}^F t) = (\alpha(t))^q,$$

so the fixed points of τ are the elements of $X(\mathbf{T})$ which are defined over $\mathbb{F}_q$.

8.1 PROPOSITION. *Let $\mathbf{T}$ be a torus of rank n; the following conditions are equivalent:*

(i) There exists an isomorphism $\mathbf{T} \xrightarrow{\sim} (\mathbb{G}_m)^n$ defined over $\mathbb{F}_q$,
(ii) The action of τ on $X(\mathbf{T})$ is trivial.

If these conditions are satisfied $\mathbf{T}$ is said to be a **split torus**.

PROOF: The fact that (i) implies (ii) follows from the easily checked equality $X(\mathbb{G}_m)^\tau = X(\mathbb{G}_m)$. Conversely, if $(a_1, \ldots, a_n)$ is a basis of the $\mathbb{Z}$-module $X(\mathbf{T})$ we have (see last sentence of the paragraph after 0.4)

$$A(\mathbf{T}) \simeq \overline{\mathbb{F}}_q[X(\mathbf{T})] \simeq \overline{\mathbb{F}}_q[a_1, \ldots, a_n].$$

But, as $X(\mathbf{T}) = X(\mathbf{T})^\tau$, all elements of $X(\mathbf{T})$ are defined over $\mathbb{F}_q$ and $T_i \mapsto a_i$ defines a rational isomorphism

$$A(\mathbb{G}_m^n) \xrightarrow{\sim} \overline{\mathbb{F}}_q[T_1, T_1^{-1}, \ldots, T_n, T_n^{-1}] \to A(\mathbf{T}).$$

■

Note that as $a_1, \ldots, a_n$ in the above proof are always defined over some finite extension of $\mathbb{F}_q$, any torus defined over $\mathbb{F}_q$ becomes split over such an extension whence we also get the fact that the order of τ is finite; the torus $\mathbf{T}$ is split over $\mathbb{F}_{q^n}$ if and only if n is multiple of the order of τ.

8.2 PROPOSITION. *Let* $\mathbf{T}$ *be a torus defined over* $\mathbb{F}_q$*; then it contains a maximum split subtorus* $\mathbf{T}_d$ *and the rank of this subtorus is equal to the rank of* $X(\mathbf{T})^\tau$.

PROOF: By 8.1 (ii), the image of a split torus by a morphism defined over $\mathbb{F}_q$ is split (as the character group of the image is a subgroup of the character group of the initial torus). This proves that the product of two split subtori of $\mathbf{T}$ is again split. So $\mathbf{T}_d$ exists.

The group $Y(\mathbf{T}_d)$ can be identified with a subgroup of $Y(\mathbf{T})$ fixed by τ, so the rank of $\mathbf{T}_d$ is not greater than that of $Y(\mathbf{T})^\tau$. But the subtorus of $\mathbf{T}$ generated by the images of $Y(\mathbf{T})^\tau$ is split of rank equal to that of $Y(\mathbf{T})^\tau$, whence the reverse inequality. Since the ranks of $X(\mathbf{T})^\tau$ and of $Y(\mathbf{T})^\tau$ are equal, because $X(\mathbf{T})$ and $Y(\mathbf{T})$ together with the τ actions are dual to each other, we get the result. ■

8.3 DEFINITION.

(i) *We call the* $\mathbb{F}_q$**-rank** *of a torus the rank of its maximum split subtorus.*

(ii) *A rational maximal torus of an algebraic group* $\mathbf{G}$ *defined over* $\mathbb{F}_q$ *is said to be* **quasi-split** *if it is contained in some rational Borel subgroup of* $\mathbf{G}$.

(iii) *We call the* $\mathbb{F}_q$**-rank** *of an algebraic group* $\mathbf{G}$*, defined over* $\mathbb{F}_q$*, the* $\mathbb{F}_q$*-rank of a quasi-split maximal torus of* $\mathbf{G}$.

With that definition, if $E = X(\mathbf{T}) \otimes \mathbb{R}$, we have $\operatorname{rank}(\mathbf{T}) = \dim E$ and, by 8.2, $\mathbb{F}_q\text{-rank}(\mathbf{T}) = \dim(E^\tau)$. Note that, as all rational Borel subgroups are rationally conjugate and all rational tori in a Borel subgroup are rationally conjugate (see 3.15), definition (iii) makes sense. Note also that the terminology is consistent as, by 8.1 (ii), any split maximal torus $\mathbf{T}$ is quasi-split; indeed there is a bijection between the Borel subgroups containing $\mathbf{T}$ and the bases of the root system of $\mathbf{G}$ relative to $\mathbf{T}$ (see 0.31 (v)), and, as τ acts trivially on $X(\mathbf{T})$, any Borel subgroup containing $\mathbf{T}$ is fixed by the Frobenius endomorphism.

8.4 DEFINITION. *We say that a reductive group* $\mathbf{G}$ *is* **split** *if a (any) quasi-split torus of* $\mathbf{G}$ *is split. Non-split groups are sometimes called* **twisted** *groups.*

Note that the $\mathbb{F}_q$-rank of a group is clearly equal to that of its reductive quotient (as tori map isomorphically in that quotient), *i.e.*, we have $\mathbb{F}_q\text{-rank}(\mathbf{G}) = \mathbb{F}_q\text{-rank}(\mathbf{G}/R_u(\mathbf{G}))$. The following lemma relates the $\mathbb{F}_q$-rank of a group to that of its semi-simple quotient.

8.5 LEMMA. *With the notation of 8.3 (iii), we have*

$$\mathbb{F}_q\text{-rank}(\mathbf{G}) = \mathbb{F}_q\text{-rank}(\mathbf{G}/R\mathbf{G}) + \mathbb{F}_q\text{-rank}(R\mathbf{G}).$$

PROOF: From the above remarks we may assume that $\mathbf{G}$ is reductive, in which case $R(\mathbf{G})$ is a torus. As the functor X is exact (see 0.21), the exact sequence

$$1 \to R\mathbf{G} \to \mathbf{T} \to \mathbf{T}/R\mathbf{G} \to 1$$

(where $\mathbf{T}$ is a rational torus) gives

$$0 \leftarrow X(R\mathbf{G}) \leftarrow X(\mathbf{T}) \leftarrow X(\mathbf{T}/R\mathbf{G}) \leftarrow 0.$$

When we take its tensor product with $\mathbb{R}$, this sequence remains exact, and it remains so when we take the subspaces of fixed points under τ, as τ, being of finite order, is semi-simple. If we take for $\mathbf{T}$ a quasi-split torus of $\mathbf{G}$, we get the result. ∎

8.6 DEFINITION. *We call the* **semi-simple $\mathbb{F}_q$-rank** *of* $\mathbf{G}$, *denoted by* $r(\mathbf{G})$, *the* $\mathbb{F}_q$*-rank of* $\mathbf{G}/R\mathbf{G}$.

8.7 LEMMA. *Let* $\mathbf{T}$ *be any rational maximal torus of a reductive group* $\mathbf{G}$, *and let* $E = X(\mathbf{T}) \otimes \mathbb{R}$; *if* $\Phi^\vee$ *is the set of coroots of* $\mathbf{G}$ *relative to* $\mathbf{T}$, *we have* $\mathbb{F}_q\text{-rank}(R\mathbf{G}) = \dim(<\Phi^\vee>^\perp \cap E^\tau)$.

PROOF: Assume that $\mathbf{T}$ is of type w with respect to a quasi-split torus; then $\mathbf{T}$ together with the action of F can be identified with the quasi-split torus with the action of wF (see 3.24); so, as the Weyl group acts trivially on $<\Phi^\vee>^\perp$, the action of τ on $<\Phi^\vee>^\perp$ does not depend on the type of $\mathbf{T}$ and we may assume that $\mathbf{T}$ is quasi-split. The lemma will be a consequence of the isomorphism $X(R\mathbf{G}) \otimes \mathbb{R} \xrightarrow{\sim} <\Phi^\vee>^\perp$. To prove it we consider the second exact sequence of 8.5 tensored by $\mathbb{R}$ and remark that Φ is the root system of the semi-simple group $\mathbf{G}/R\mathbf{G}$ relative to the quasi-split maximal torus $\mathbf{T}/R(\mathbf{G})$, so spans $X(\mathbf{T}/R(\mathbf{G})) \otimes \mathbb{R}$ (see 0.37 (iii)). ∎

We can now define the duality operator.

8.8 DEFINITION. *Let* $\mathbf{B}$ *a rational Borel subgroup of a connected reductive algebraic group* $\mathbf{G}$ *defined over* $\mathbb{F}_q$. *By* **duality** *we mean the operator* $D_\mathbf{G}$ *on* $\mathcal{C}(\mathbf{G}^F)$ *defined by*

$$D_\mathbf{G} = \sum_{\mathbf{P} \supset \mathbf{B}} (-1)^{r(\mathbf{P})} R_\mathbf{L}^\mathbf{G} \circ {}^*R_\mathbf{L}^\mathbf{G}$$

where the summation is over the set of rational parabolic subgroups of $\mathbf{G}$ *containing* $\mathbf{B}$ *and where* $\mathbf{L}$ *denotes an arbitrarily chosen rational Levi subgroup of* $\mathbf{P}$.

8.9 REMARKS. *This definition is coherent because of the two following properties:*

(i) *The operator* $R_{\mathbf{L}}^{\mathbf{G}} \circ {}^*R_{\mathbf{L}}^{\mathbf{G}}$ *depends only on the parabolic subgroup* $\mathbf{P}$ *and not on the chosen Levi subgroup.*

(ii) *The operator* $D_{\mathbf{G}}$ *does not depend on the rational Borel subgroup* $\mathbf{B}$.

Property (i) comes from the fact that two rational Levi subgroups of $\mathbf{P}$ are rationally conjugate and from the equality $R_{\mathbf{L}}^{\mathbf{G}} \circ \operatorname{ad} x^{-1} = R_{^x\mathbf{L}}^{\mathbf{G}}$ (along with its analogue for ${}^*R_{\mathbf{L}}^{\mathbf{G}}$). We could also deduce it from the equality $R_{\mathbf{L}}^{\mathbf{G}} \circ {}^*R_{\mathbf{L}}^{\mathbf{G}}(\chi) = \operatorname{Ind}_{\mathbf{P}^F}^{\mathbf{G}^F}(\chi')$ where we have put $\chi'(p) = |\mathbf{U}^F|^{-1} \sum_{u \in \mathbf{U}^F} \chi(pu)$ for $p \in \mathbf{P}^F$, with $\mathbf{U}$ denoting the unipotent radical of $\mathbf{P}$.

Property (ii) is clear as all rational Borel subgroups of $\mathbf{G}$ are rationally conjugate.

8.10 PROPOSITION. *The functor* $D_{\mathbf{G}}$ *is self-adjoint.*

This is a straightforward consequence of the definition of $D_{\mathbf{G}}$ and of the fact that $R_{\mathbf{L}}^{\mathbf{G}}$ and ${}^*R_{\mathbf{L}}^{\mathbf{G}}$ are adjoint to each other (see 4.6 (iii)).

8.11 THEOREM (C. CURTIS). *For any rational Levi subgroup* $\mathbf{L}$ *of a rational parabolic subgroup of* $\mathbf{G}$ *we have*

$$D_{\mathbf{G}} \circ R_{\mathbf{L}}^{\mathbf{G}} = R_{\mathbf{L}}^{\mathbf{G}} \circ D_{\mathbf{L}}.$$

The proof we shall give is valid even when the parabolic subgroup is not rational, in which case we have to replace $R_{\mathbf{L}}^{\mathbf{G}}$ with the Lusztig functor which is its generalization and is again denoted by $R_{\mathbf{L}}^{\mathbf{G}}$ (this functor will be extensively studied later, starting from chapter 11) and the formula has to be written

$$\varepsilon_{\mathbf{G}} D_{\mathbf{G}} \circ R_{\mathbf{L}}^{\mathbf{G}} = \varepsilon_{\mathbf{L}} R_{\mathbf{L}}^{\mathbf{G}} \circ D_{\mathbf{L}}$$

where for any algebraic group $\mathbf{G}$, we have put $\varepsilon_{\mathbf{G}} = (-1)^{\mathbb{F}_q\text{-rank}(\mathbf{G})}$. In the case of a rational parabolic subgroup this formula reduces to the previous one because $\mathbf{L}$ then contains a quasi-split torus of $\mathbf{G}$, so that $\mathbb{F}_q$-rank$(\mathbf{G}) = \mathbb{F}_q$-rank$(\mathbf{L})$.

The only properties of $R_{\mathbf{L}}^{\mathbf{G}}$ that we shall use in the proof are transitivity, the formula $R_{\mathbf{L}}^{\mathbf{G}} = R_{^x\mathbf{L}}^{\mathbf{G}} \circ \operatorname{ad} x$ (for $x \in \mathbf{G}^F$) and the "Mackey formula"; these formulae are true for the Lusztig functor, with some restriction for the Mackey formula (see 11.6). The reader will check that in the remainder of the book we shall use 8.11 for the Lusztig functor only in those cases where the Mackey formula has been proved.

PROOF: By the Mackey formula, using the equality $(\mathbf{M}^F\backslash\mathcal{S}(\mathbf{M},\mathbf{L})^F/\mathbf{L}^F)^{-1} = \mathbf{L}^F\backslash\mathcal{S}(\mathbf{L},\mathbf{M})^F/\mathbf{M}^F$ we get

$$\begin{aligned} D_{\mathbf{G}} \circ R_{\mathbf{L}}^{\mathbf{G}} &= \sum_{\mathbf{Q}\supset\mathbf{B}} (-1)^{r(\mathbf{Q})} R_{\mathbf{M}}^{\mathbf{G}}\,{}^*R_{\mathbf{M}}^{\mathbf{G}} R_{\mathbf{L}}^{\mathbf{G}} \\ &= \sum_{\mathbf{Q}\supset\mathbf{B}} (-1)^{r(\mathbf{Q})} \sum_{x\in\mathbf{L}^F\backslash\mathcal{S}(\mathbf{L},\mathbf{M})^F/\mathbf{M}^F} R_{\mathbf{M}}^{\mathbf{G}} \operatorname{ad} x^{-1} R_{{}^x\mathbf{M}\cap\mathbf{L}}^{{}^x\mathbf{M}}\,{}^*R_{{}^x\mathbf{M}\cap\mathbf{L}}^{\mathbf{L}}. \end{aligned}$$

We then use the equalities $R_{\mathbf{M}}^{\mathbf{G}} \operatorname{ad} x^{-1} = R_{{}^x\mathbf{M}}^{\mathbf{G}}$ and $R_{{}^x\mathbf{M}}^{\mathbf{G}} R_{{}^x\mathbf{M}\cap\mathbf{L}}^{{}^x\mathbf{M}} = R_{{}^x\mathbf{M}\cap\mathbf{L}}^{\mathbf{G}} = R_{\mathbf{L}}^{\mathbf{G}} R_{{}^x\mathbf{M}\cap\mathbf{L}}^{\mathbf{L}}$ to get

$$D_{\mathbf{G}} \circ R_{\mathbf{L}}^{\mathbf{G}} = R_{\mathbf{L}}^{\mathbf{G}} \circ \Big(\sum_{\mathbf{Q}\supset\mathbf{B}} (-1)^{r(\mathbf{Q})} \sum_{x\in\mathbf{L}^F\backslash\mathcal{S}(\mathbf{L},\mathbf{M})^F/\mathbf{M}^F} R_{{}^x\mathbf{M}\cap\mathbf{L}}^{\mathbf{L}}\,{}^*R_{{}^x\mathbf{M}\cap\mathbf{L}}^{\mathbf{L}}\Big).$$

We now transform the right-hand side of the above equality. Let $\mathbf{B}_0$ be a fixed rational Borel subgroup of $\mathbf{L}$. For each parabolic subgroup $\mathbf{Q} \supset \mathbf{B}$ of $\mathbf{G}$ we define a bijection $\mathcal{P}$ from $\mathbf{L}^F\backslash\mathcal{S}(\mathbf{L},\mathbf{M})^F/\mathbf{M}^F$ to the set of parabolic subgroups of $\mathbf{G}$ which are $\mathbf{G}^F$-conjugate to $\mathbf{Q}$ and contain $\mathbf{B}_0$. If $x \in \mathcal{S}(\mathbf{L},\mathbf{M})^F$, as ${}^x\mathbf{Q}$ and $\mathbf{L}$ contain a common maximal torus, the group ${}^x\mathbf{Q} \cap \mathbf{L}$ is a parabolic subgroup of $\mathbf{L}$, see 2.1 (i). As it is rational this parabolic subgroup contains a rational Borel subgroup of $\mathbf{L}$ and such a Borel subgroup is conjugate by an element $l \in \mathbf{L}^F$ to $\mathbf{B}_0$. Then we have $\overline{lx} = \overline{x}$, where $\overline{x}$ denotes the image of x in $\mathbf{L}^F\backslash\mathcal{S}(\mathbf{L},\mathbf{M})^F/\mathbf{M}^F$, and ${}^{lx}\mathbf{Q} \supset \mathbf{B}_0$. We put $\mathcal{P}(\overline{x}) = {}^{lx}\mathbf{Q}$. This map is well-defined (*i.e.*, ${}^{lx}\mathbf{Q}$ does not depend on the chosen l) since two conjugate parabolic subgroups of $\mathbf{L}$ containing the same Borel subgroup are equal (see 0.12 (ii)).

The map $\mathcal{P}$ is onto. Indeed, let ${}^x\mathbf{Q}$ with $x \in \mathbf{G}^F$ be a parabolic subgroup in $\mathbf{G}^F$ containing $\mathbf{B}_0$, let $\mathbf{T}_0$ be a maximal torus of $\mathbf{B}_0$ and let $\mathbf{M}'$ be the unique Levi subgroup of ${}^x\mathbf{Q}$ containing $\mathbf{T}_0$ (see 1.17). Then $\mathbf{M}'$ and ${}^x\mathbf{M}$ are conjugate under ${}^x\mathbf{Q}^F$, *i.e.*, there exists $q \in \mathbf{Q}^F$ such that $\mathbf{M}' = {}^{xq}\mathbf{M}$. But then ${}^{xq}\mathbf{M} \cap \mathbf{L} \supset \mathbf{T}_0$, so xq is in $\mathcal{S}(\mathbf{L},\mathbf{M})^F$ and satisfies $\mathcal{P}(\overline{xq}) = {}^{xq}\mathbf{Q} = {}^x\mathbf{Q}$.

We now prove the injectivity of $\mathcal{P}$. If ${}^x\mathbf{Q} = {}^y\mathbf{Q}$ with x and y in $\mathcal{S}(\mathbf{L},\mathbf{M})^F$ then there exists $q \in \mathbf{Q}^F$ such that $y = xq$ (because $N_{\mathbf{G}^F}(\mathbf{Q}) = \mathbf{Q}^F$). The groups $\mathbf{L} \cap {}^x\mathbf{M}$ and $\mathbf{L} \cap {}^{xq}\mathbf{M}$ are two rational Levi subgroups of the parabolic subgroup $\mathbf{L} \cap {}^x\mathbf{Q}$ of $\mathbf{L}$, so are conjugate by an element $z \in \mathbf{L}^F \cap {}^x\mathbf{Q}$. We have $\mathbf{L} \cap {}^{zxq}\mathbf{M} = {}^z(\mathbf{L} \cap {}^{xq}\mathbf{M}) = \mathbf{L} \cap {}^x\mathbf{M}$, so ${}^x\mathbf{M}$ and ${}^{zxq}\mathbf{M}$ are two Levi subgroups of ${}^x\mathbf{Q}$ containing a common maximal torus (any maximal torus of $\mathbf{L} \cap {}^x\mathbf{M}$), so are equal; thus $zxqx^{-1} \in N_{\mathbf{G}^F}({}^x\mathbf{M}) \cap {}^x\mathbf{Q} = {}^x\mathbf{M}$, so

$y = xq \in z^{-1}x\mathbf{M} \subset \mathbf{L}x\mathbf{M}$ which implies, using the Lang-Steinberg theorem in $\mathbf{L} \cap {}^x\mathbf{M}$, that $y \in \mathbf{L}^F x \mathbf{M}^F$, whence $\overline{y} = \overline{x}$.

We have noticed in 8.9 (i) that the functor $R^{\mathbf{L}}_{{}^x\mathbf{M}\cap\mathbf{L}}{}^*R^{\mathbf{L}}_{{}^x\mathbf{M}\cap\mathbf{L}}$ depends only on ${}^x\mathbf{Q}\cap\mathbf{L}$. We denote by $f({}^x\mathbf{Q})$ this functor; we have $f({}^x\mathbf{Q}) = f(\mathcal{P}(\overline{x}))$ as $\mathcal{P}(\overline{x})$ and ${}^x\mathbf{Q}$ are conjugate under $\mathbf{L}^F$. Since $\mathcal{P}$ is bijective and $r(\mathbf{Q}) = r({}^x\mathbf{Q}) = r(\mathcal{P}(\overline{x}))$, we get

$$D_{\mathbf{G}} \circ R^{\mathbf{G}}_{\mathbf{L}} = R^{\mathbf{G}}_{\mathbf{L}} \circ \sum_{\mathbf{Q}'} (-1)^{r(\mathbf{Q}')} f(\mathbf{Q}')$$

where $\mathbf{Q}'$ in the summation runs over all rational parabolic subgroups of $\mathbf{G}$ containing $\mathbf{B}_0$. The summation in the right-hand side can be written

$$\sum_{\mathbf{Q}_0 \supset \mathbf{B}_0} \Big(\sum_{\{\mathbf{Q}' \mid \mathbf{Q}' \cap \mathbf{L} = \mathbf{Q}_0\}} (-1)^{r(\mathbf{Q}')} \Big) R^{\mathbf{L}}_{\mathbf{M}_0} \circ {}^*R^{\mathbf{L}}_{\mathbf{M}_0}$$

where $\mathbf{Q}_0$ runs over rational parabolic subgroups of $\mathbf{L}$ containing $\mathbf{B}_0$, where $\mathbf{M}_0$ denotes a rational Levi subgroup of $\mathbf{Q}_0$, and $\mathbf{Q}'$ runs over rational parabolic subgroups of $\mathbf{G}$ such that $\mathbf{Q}' \cap \mathbf{L} = \mathbf{Q}_0$. Theorem 8.11 is thus a consequence of the following result.

8.12 THEOREM. *Let $\mathbf{H}$ be a reductive rational subgroup of maximal rank of $\mathbf{G}$, and let $\mathbf{Q}_0$ be a rational parabolic subgroup of $\mathbf{H}$. Then we have*

$$\varepsilon_{\mathbf{G}} \sum_{\{\mathbf{Q} \mid \mathbf{Q} \cap \mathbf{H} = \mathbf{Q}_0\}} (-1)^{r(\mathbf{Q})} = (-1)^{r(\mathbf{Q}_0)} \varepsilon_{\mathbf{H}}$$

where $\mathbf{Q}$ in the summation runs over rational parabolic subgroups of $\mathbf{G}$ such that $\mathbf{Q} \cap \mathbf{H} = \mathbf{Q}_0$.

PROOF: Since $\mathbf{Q}$ is rational it has the same $\mathbb{F}_q$-rank as $\mathbf{G}$, so $\mathbb{F}_q$-rank$(\mathbf{G}) - r(\mathbf{Q}) = \mathbb{F}_q$-rank$(R(\mathbf{Q}))$, see 8.5. We fix a maximal torus $\mathbf{T}$ of $\mathbf{Q}_0$; using 8.7 we shall translate the statement in $E = X(\mathbf{T}) \otimes \mathbb{R}$. Let $\mathbf{M}$ be the Levi subgroup of $\mathbf{Q}$ which contains $\mathbf{T}$; we have $\mathbb{F}_q$-rank$(R(\mathbf{Q})) = \mathbb{F}_q$-rank$(R(\mathbf{M}))$. Let $\Phi^\vee_{\mathbf{G}}$, $\Phi^\vee_{\mathbf{Q}}$ and $\Phi^\vee_{\mathbf{M}}$ be respectively the sets of coroots of $\mathbf{G}$, $\mathbf{Q}$ and $\mathbf{M}$ with respect to $\mathbf{T}$; we associate to $\mathbf{G}$ the set of hyperplanes $\mathcal{H}_{\mathbf{G}}$ in E consisting of $\{\alpha^\perp \mid \alpha \in \Phi^\vee_{\mathbf{G}}\}$, and to $\mathbf{Q}$ the facet of this system defined by

$$\mathcal{F}_{\mathbf{Q}} = \{\, x \in E \mid \langle\, x, \alpha \,\rangle = 0 \text{ for } \alpha \in \Phi^\vee_{\mathbf{M}} \text{ and } \langle\, x, \alpha \,\rangle > 0 \text{ for } \alpha \in \Phi^\vee_{\mathbf{Q}} - \Phi^\vee_{\mathbf{M}} \,\}$$

(it is a facet as $\Phi^\vee_{\mathbf{Q}} \cup -\Phi^\vee_{\mathbf{Q}} = \Phi^\vee_{\mathbf{G}}$). In the same way we associate to $\mathbf{Q}_0$ a facet $\mathcal{F}_{\mathbf{Q}_0}$ of the set of hyperplanes $\mathcal{H}_{\mathbf{H}} = \{\alpha^\perp \mid \alpha \in \Phi^\vee_{\mathbf{H}}\}$ where $\Phi^\vee_{\mathbf{H}}$ is the set of coroots of $\mathbf{H}$ relative to $\mathbf{T}$. Note that $\mathcal{F}_{\mathbf{Q}}$ is open in its support $<\Phi^\vee_{\mathbf{M}}>^\perp$ and that $\mathcal{F}_{\mathbf{Q}} \cap E^\tau \neq \emptyset$ as ${}^\tau\mathcal{F}_{\mathbf{Q}} = \mathcal{F}_{\mathbf{Q}}$, so that $\dim(\mathcal{F}_{\mathbf{Q}} \cap E^\tau) = \dim(<\Phi^\vee_{\mathbf{M}}>^\perp \cap E^\tau) = \mathbb{F}_q$-rank$(R\mathbf{M}))$. The formula we have to prove can now be written

$$\sum_{\{\mathbf{Q} \mid \mathbf{Q} \cap \mathbf{H} = \mathbf{Q}_0\}} (-1)^{\dim(\mathcal{F}_{\mathbf{Q}} \cap E^\tau)} = (-1)^{\dim(\mathcal{F}_{\mathbf{Q}_0} \cap E^\tau)}.$$

But any facet of $\mathcal{H}_{\mathbf{G}}$ is equal to some $\mathcal{F}_{\mathbf{Q}}$ when $\mathbf{Q}$ runs over the parabolic subgroups which contain $\mathbf{T}$ (since, for $x \in E$, the set $\{\,\alpha \in \Phi^\vee_{\mathbf{G}} \mid \langle\, x, \alpha\,\rangle \geq 0\,\}$ defines a parabolic subset of $\Phi^\vee_{\mathbf{G}}$ such that x is in the corresponding facet; see 1.20 (ii)), so the same property is true for the set of hyperplanes which is the trace in E^τ of $\mathcal{H}_{\mathbf{G}}$. Moreover

$$\mathbf{Q} \cap \mathbf{H} = \mathbf{Q}_0 \Leftrightarrow \mathcal{F}_{\mathbf{Q}} \subset \mathcal{F}_{\mathbf{Q}_0},$$

so we can rewrite the formula as

$$\sum_{\mathcal{F} \subset \mathcal{F}_{\mathbf{Q}_0} \cap E^\tau} (-1)^{\dim \mathcal{F}} = (-1)^{\dim \mathcal{F}_{\mathbf{Q}_0} \cap E^\tau}.$$

But this last formula is well-known: it expresses the fact that the Euler characteristic of the convex open set $\mathcal{F}_{\mathbf{Q}_0} \cap E^\tau$ can be computed using the subdivision into facets defined by $\mathcal{H}_{\mathbf{G}} \cap E^\tau$. ■

8.13 COROLLARY (OF 8.11). $^*R^{\mathbf{G}}_{\mathbf{L}} \circ D_{\mathbf{G}} = D_{\mathbf{L}} \circ {}^*R^{\mathbf{G}}_{\mathbf{L}}$.

PROOF: This is just the adjoint formula. ■

8.14 COROLLARY. $D_{\mathbf{G}} \circ D_{\mathbf{G}}$ *is the identity functor.*

PROOF: Let $\mathbf{B}$ be a rational Borel subgroup of $\mathbf{G}$ and let $\mathbf{T}$ be a rational maximal torus of $\mathbf{B}$. The Borel subgroup $\mathbf{B}$ determines a τ-stable basis Π of the root system Φ of $\mathbf{G}$ relative to $\mathbf{T}$. The rational parabolic subgroups of $\mathbf{G}$ which contain $\mathbf{B}$ correspond one-to-one to the τ-stable subsets of Π, *i.e.*, to the subsets of the set $\overline{\Pi}$ of τ-orbits in Π. If $\mathbf{P}_I$ denotes the parabolic subgroup corresponding to $I \subset \overline{\Pi}$, then $r(\mathbf{P}_I) = |I|$. Indeed, for $\mathbf{L}_I$ a rational Levi subgroup of $\mathbf{P}_I$ containing $\mathbf{T}$, we saw in the proof of 8.12 that $r(\mathbf{P}_I) = \mathbb{F}_q\text{-rank}(\mathbf{G}) - \mathbb{F}_q\text{-rank}(R\mathbf{L}_I)$; but $\mathbb{F}_q\text{-rank}(\mathbf{G}) = \dim(E^\tau)$ and $\mathbb{F}_q\text{-rank}(R\mathbf{L}_I) = \dim(<\Phi^\vee_{\mathbf{L}_I}>^\perp \cap E^\tau)$, so $r(\mathbf{P}_I) = \dim(<\Phi_{\mathbf{L}_I}> \cap E^\tau) = |I|$.

Thus we can write $D_{\mathbf{G}} = \sum_{I \subset \overline{\Pi}} (-1)^{|I|} R^{\mathbf{G}}_{\mathbf{L}_I} {}^*R^{\mathbf{G}}_{\mathbf{L}_I}$, whence

$$\begin{aligned} D_{\mathbf{G}} \circ D_{\mathbf{G}} &= \sum_{I \subset \overline{\Pi}} (-1)^{|I|} R^{\mathbf{G}}_{\mathbf{L}_I} \circ {}^*R^{\mathbf{G}}_{\mathbf{L}_I} \circ D_{\mathbf{G}} = \sum_{I \subset \overline{\Pi}} (-1)^{|I|} R^{\mathbf{G}}_{\mathbf{L}_I} \circ D_{\mathbf{L}_I} \circ {}^*R^{\mathbf{G}}_{\mathbf{L}_I} \\ &= \sum_{I \subset \overline{\Pi}} (-1)^{|I|} \sum_{K \subset I} (-1)^{|K|} R^{\mathbf{G}}_{\mathbf{L}_I} \circ R^{\mathbf{L}_I}_{\mathbf{L}_K} \circ {}^*R^{\mathbf{L}_I}_{\mathbf{L}_K} \circ {}^*R^{\mathbf{G}}_{\mathbf{L}_I} \\ &= \sum_{K \subset \overline{\Pi}} (-1)^{|K|} \Big(\sum_{I \supset K} (-1)^{|I|}\Big) R^{\mathbf{G}}_{\mathbf{L}_K} \circ {}^*R^{\mathbf{G}}_{\mathbf{L}_K} = \mathrm{Id}; \end{aligned}$$

the last equality uses the fact that $\sum_{I \supset K} (-1)^{|I|} = 0$ if $K \neq \overline{\Pi}$. ■

The following result shows that the dual of an irreducible character is an irreducible character up to a sign determined by the Harish-Chandra series of the character.

8.15 COROLLARY. *Let $\delta \in \operatorname{Irr}(\mathbf{L}^F)$ be a cuspidal character and let $\chi \in \operatorname{Irr}(\mathbf{G}^F)$ be such that $\langle \chi, R_{\mathbf{L}}^{\mathbf{G}}\delta \rangle_{\mathbf{G}^F} \neq 0$. Then $(-1)^{r(\mathbf{L})} D_{\mathbf{G}}(\chi) \in \operatorname{Irr}(\mathbf{G}^F)$.*

PROOF: We have $\langle \chi, \chi \rangle_{\mathbf{G}^F} = \langle \chi, D_{\mathbf{G}} \circ D_{\mathbf{G}}\chi \rangle_{\mathbf{G}^F} = \langle D_{\mathbf{G}}\chi, D_{\mathbf{G}}\chi \rangle_{\mathbf{G}^F}$ (see 8.10), so $D_{\mathbf{G}}(\chi)$ is an irreducible character up to sign. As δ is cuspidal, all summands in the formula for $D_{\mathbf{L}}(\delta)$ are zero except one of them and we have $D_{\mathbf{L}}(\delta) = (-1)^{r(\mathbf{L})}\delta$. So, by 8.10, 8.11 and this remark, we have

$$\begin{aligned}(-1)^{r(\mathbf{L})}\langle D_{\mathbf{G}}\chi, R_{\mathbf{L}}^{\mathbf{G}}\delta \rangle_{\mathbf{G}^F} &= (-1)^{r(\mathbf{L})}\langle \chi, D_{\mathbf{G}} R_{\mathbf{L}}^{\mathbf{G}}\delta \rangle_{\mathbf{G}^F} \\ &= (-1)^{r(\mathbf{L})}\langle \chi, R_{\mathbf{L}}^{\mathbf{G}} D_{\mathbf{L}}\delta \rangle_{\mathbf{G}^F} = \langle \chi, R_{\mathbf{L}}^{\mathbf{G}}\delta \rangle_{\mathbf{G}^F} > 0,\end{aligned}$$

whence the result. ■

8.16 COROLLARY (OF 8.12). *If s is the semi-simple part of an element $x \in \mathbf{G}^F$, then*

$$\varepsilon_{\mathbf{G}}(D_{\mathbf{G}}\chi)(x) = \varepsilon_{C^\circ_{\mathbf{G}}(s)}(D_{C^\circ_{\mathbf{G}}(s)} \circ \operatorname{Res}^{\mathbf{G}}_{C^\circ_{\mathbf{G}}(s)} \chi)(x).$$

PROOF: If $\mathbf{L}$ is a rational Levi subgroup of a rational parabolic subgroup $\mathbf{P}$ of $\mathbf{G}$, we have (see remarks after 8.9) $R_{\mathbf{L}}^{\mathbf{G}} \circ {}^*R_{\mathbf{L}}^{\mathbf{G}}\chi = \operatorname{Ind}_{\mathbf{P}^F}^{\mathbf{G}^F}(\chi')$, if $\chi'(p) = {}^*R_{\mathbf{L}}^{\mathbf{G}}(\chi)(\overline{p})$ where $\overline{p}$ is the image of p in $\mathbf{L}$, whence

$$(R_{\mathbf{L}}^{\mathbf{G}} \circ {}^*R_{\mathbf{L}}^{\mathbf{G}}\chi)(x) = |\mathbf{P}^F|^{-1} \sum_{\{g \in \mathbf{G}^F \mid {}^g\mathbf{P} \ni x\}} ({}^*R_{\mathbf{L}}^{\mathbf{G}}\chi)(\overline{{}^{g^{-1}}x}) = \sum_{\{\mathbf{P}' \underset{\mathbf{G}^F}{\sim} \mathbf{P} \mid \mathbf{P}' \ni x\}} ({}^*R_{\mathbf{L}'}^{\mathbf{G}}\chi)(\overline{x})$$

where $\mathbf{P}'$ in the last summation denotes a rational parabolic subgroup of $\mathbf{G}$, where $\mathbf{L}'$ is a rational Levi subgroup of $\mathbf{P}'$, and $\overline{x}$ denotes the image of x in $\mathbf{L}'$. As any rational parabolic subgroup of $\mathbf{G}$ is $\mathbf{G}^F$-conjugate to some $\mathbf{P} \supset \mathbf{B}$, we have

$$(D_{\mathbf{G}}\chi)(x) = \sum_{\mathbf{P} \supset \mathbf{B}} (-1)^{r(\mathbf{P})} \sum_{\{\mathbf{P}' \underset{\mathbf{G}^F}{\sim} \mathbf{P} \mid \mathbf{P}' \ni x\}} ({}^*R_{\mathbf{L}'}^{\mathbf{G}}\chi)(\overline{x}) = \sum_{\mathbf{P}' \ni x} (-1)^{r(\mathbf{P}')}({}^*R_{\mathbf{L}'}^{\mathbf{G}}\chi)(\overline{x}). \tag{1}$$

By 7.6 we then get

$$\begin{aligned}(D_{\mathbf{G}}\chi)(x) &= \sum_{\mathbf{P} \ni x} (-1)^{r(\mathbf{P})}({}^*R_{\mathbf{L} \cap C^\circ_{\mathbf{G}}(s)}^{C^\circ_{\mathbf{G}}(s)}\chi)(\overline{x}) \\ &= \sum_{\mathbf{P}'} \Big(\sum_{\{\mathbf{P} \mid \mathbf{P} \cap C^\circ_{\mathbf{G}}(s) = \mathbf{P}'\}} (-1)^{r(\mathbf{P})}\Big)({}^*R_{\mathbf{L}'}^{C^\circ_{\mathbf{G}}(s)}\chi)(\overline{x})\end{aligned}$$

where $\mathbf{P}'$ runs over the set of rational parabolic subgroups of $C^\circ_{\mathbf{G}}(s)$ and where $\mathbf{L}'$ is any rational Levi subgroup of $\mathbf{P}'$. If we now apply 8.12 with $\mathbf{H} = C^\circ_{\mathbf{G}}(s)$ and compare with the equality (1) applied in $C^\circ_{\mathbf{G}}(s)$, we get the result. ■

References

The duality operator $D_{\mathbf{G}}$ was introduced simultaneously by Curtis [Cu], Alvis [Duality in the character ring of a finite Chevalley group, *Proc. Symp. in Pure Math.*, **37** (1980), 353–357], Kawanaka [Fourier transforms of nilpotently supported invariant functions on a simple Lie algebra over a finite field, *Inventiones Math.*, **69** (1982), 411–435, §2] and Lusztig. Deligne and Lusztig studied it in a geometric way in [DL2]; see also [DM1] which we follow here, in particular for the proof of 8.11. The notions of $\mathbb{F}_q$-rank and $\mathbb{F}_q$-semi-simple rank can be found in [BT], *e.g.*, statements 4.23 and 5.3.

9. THE STEINBERG CHARACTER

In this chapter we use duality to define and study the famous "Steinberg character" which was originally defined by Steinberg in [R. STEINBERG Prime power representations of finite linear groups I, II, *Canad. J. Math.*, **8** (1956), 580–591; **9** (1957), 347–351].

9.1 DEFINITION. *The irreducible character* $\mathrm{St}_{\mathbf{G}} = D_{\mathbf{G}}(\mathrm{Id}_{\mathbf{G}})$ *where* $\mathrm{Id}_{\mathbf{G}}$ *is the trivial character of* $\mathbf{G}^F$ *is called* **the Steinberg character** *of* $\mathbf{G}^F$.

Note that $\mathrm{St}_{\mathbf{G}}$ is a true character (and not the opposite of a character) as, by 7.4, for any rational Levi subgroup $\mathbf{L}$ of a rational parabolic subgroup we have ${}^*R_{\mathbf{L}}^{\mathbf{G}}(\mathrm{Id}_{\mathbf{G}}) = \mathrm{Id}_{\mathbf{L}}$, thus in particular $\langle \mathrm{Id}_{\mathbf{G}}, R_{\mathbf{T}}^{\mathbf{G}}(\mathrm{Id}_{\mathbf{T}})\rangle_{\mathbf{G}^F} \neq 0$ where $\mathbf{T}$ is a quasi-split torus of $\mathbf{G}$, and as $r(\mathbf{T}) = 0$ we get the result by 8.15. Note also that to avoid cumbersome notation we write $\mathrm{Id}_{\mathbf{G}}$ and $\mathrm{St}_{\mathbf{G}}$ instead of $\mathrm{Id}_{\mathbf{G}^F}$ and $\mathrm{St}_{\mathbf{G}^F}$; we shall later use the same kind of notation for the regular representation.

As the duality and the Harish-Chandra restriction commute we get

$$ {}^*R_{\mathbf{L}}^{\mathbf{G}}\mathrm{St}_{\mathbf{G}} = {}^*R_{\mathbf{L}}^{\mathbf{G}} D_{\mathbf{G}}(\mathrm{Id}_{\mathbf{G}}) = D_{\mathbf{L}}{}^*R_{\mathbf{L}}^{\mathbf{G}}(\mathrm{Id}_{\mathbf{G}}) = D_{\mathbf{L}}(\mathrm{Id}_{\mathbf{L}}) = \mathrm{St}_{\mathbf{L}}\,. $$

In the case of a torus we have the following more precise result.

9.2 LEMMA. *Let* $\mathbf{T}$ *be a rational maximal torus of a rational Borel subgroup* $\mathbf{B}$ *of* $\mathbf{G}$; *then*

$$ \mathrm{Res}_{\mathbf{B}^F}^{\mathbf{G}^F} \mathrm{St}_{\mathbf{G}} = \mathrm{Ind}_{\mathbf{T}^F}^{\mathbf{B}^F} \mathrm{Id}_{\mathbf{T}}\,. $$

PROOF: Using the definitions of $\mathrm{St}_{\mathbf{G}}$ and of $D_{\mathbf{G}}$ we get

$$ \mathrm{St}_{\mathbf{G}} = \sum_{I\subset\overline{\Pi}} (-1)^{|I|} \mathrm{Ind}_{\mathbf{P}_I^F}^{\mathbf{G}^F} \mathrm{Id}_{\mathbf{P}_I} $$

(the notation is the same as in the proof of 8.14). So we have

$$ \begin{aligned} \mathrm{Res}_{\mathbf{B}^F}^{\mathbf{G}^F}(\mathrm{St}_{\mathbf{G}}) &= \sum_{I\subset\overline{\Pi}} (-1)^{|I|} \mathrm{Res}_{\mathbf{B}^F}^{\mathbf{G}^F} \mathrm{Ind}_{\mathbf{P}_I^F}^{\mathbf{G}^F} \mathrm{Id}_{\mathbf{P}_I} \\ &= \sum_{I\subset\overline{\Pi}} (-1)^{|I|} \sum_{w\in Z_I} \mathrm{Ind}_{\mathbf{B}^F\cap{}^w\mathbf{P}_I}^{\mathbf{B}^F} \mathrm{Res}_{\mathbf{B}^F\cap{}^w\mathbf{P}_I}^{\mathbf{P}_I^F} \mathrm{Id}_{\mathbf{P}_I}, \end{aligned} $$

the last equality following from the Mackey formula for induction and restriction, where we have denoted by Z_I the set of rational reduced-I' elements of the Weyl group W of $\mathbf{T}$ if $I \subset \overline{\Pi}$ is the image of the F-stable subset $I' \subset \Pi$ (the set Z_I is a set of representatives for the rational double cosets $\mathbf{B}\backslash\mathbf{G}/\mathbf{P}_I$ and is also a set of representatives for the double cosets $\mathbf{B}^F\backslash\mathbf{G}^F/\mathbf{P}_I^F$; see 5.5 and 5.6). But we have $\mathbf{B} \cap {}^w\mathbf{P}_I = \mathbf{B} \cap {}^w\mathbf{B}$. Indeed, as for any $v \in W_{I'}$ we have $l(w) + l(v) = l(wv)$, so for any $v \in W_{I'}$ we have $w\mathbf{B}v\mathbf{B} \subset \mathbf{B}wv\mathbf{B}$; whence $\mathbf{B}w \cap w\mathbf{B}v\mathbf{B} \subset \mathbf{B}w \cap \mathbf{B}wv\mathbf{B}$, and this last intersection is empty if v is different from 1. So we have

$$\mathbf{B}w \cap w\mathbf{P}_I = \mathbf{B}w \cap w\mathbf{B}W_{I'}\mathbf{B} = \mathbf{B}w \cap \left(\coprod_{v \in W_{I'}} w\mathbf{B}v\mathbf{B}\right) = \mathbf{B}w \cap w\mathbf{B}.$$

Using this result in the formula for $\operatorname{Res}^{\mathbf{G}^F}_{\mathbf{B}^F} \operatorname{St}_{\mathbf{G}}$ and exchanging the summations give

$$\operatorname{Res}^{\mathbf{G}^F}_{\mathbf{B}^F} \operatorname{St}_{\mathbf{G}} = \sum_{w \in W} \Big(\sum_{\{I \subset \overline{\Pi} \mid w \in Z_I\}} (-1)^{|I|}\Big) \operatorname{Ind}^{\mathbf{B}^F}_{\mathbf{B}^F \cap {}^w\mathbf{B}} \operatorname{Id}_{\mathbf{B} \cap {}^w\mathbf{B}}.$$

For $w \in W^F$ we put $\Pi_w = \{ i \in \overline{\Pi} \mid \exists \alpha \in i, {}^w\alpha > 0 \}$ (if this condition is satisfied for one $\alpha \in i$, it is also satisfied for any element of i, as $w \in W^F$). The set Z_I contains $w \in W^F$ if and only if $I \subset \Pi_w$, so

$$\sum_{\{I \subset \overline{\Pi} \mid w \in Z_I\}} (-1)^{|I|} = \sum_{I \subset \Pi_w} (-1)^{|I|},$$

which is different from zero only if $\Pi_w = \emptyset$, in which case w maps any positive root to a negative one. This happens for one and only one element of W, the longest element (see, *e.g.*, [Bbk, VI Cor. 3 of 1.6]), which is rational because of its uniqueness, and in that case we have $\mathbf{B} \cap {}^w\mathbf{B} = \mathbf{T}$, whence the result. ■

9.3 COROLLARY. *We have*

$$\operatorname{St}_{\mathbf{G}}(x) = \begin{cases} \varepsilon_{\mathbf{G}}\varepsilon_{(C_{\mathbf{G}}(x)^\circ)} |C_{\mathbf{G}}(x)^{\circ F}|_p & \textit{if } x \textit{ is semi-simple,} \\ 0 & \textit{otherwise.} \end{cases}$$

PROOF: Let s be the semi-simple part of x. We have

$$\operatorname{St}_{\mathbf{G}}(x) = D_{\mathbf{G}}(\operatorname{Id}_{\mathbf{G}})(x) = \varepsilon_{\mathbf{G}}\varepsilon_{C^\circ_{\mathbf{G}}(s)}(D_{C^\circ_{\mathbf{G}}(s)}(\operatorname{Id}_{C^\circ_{\mathbf{G}}(s)}))(x) = \varepsilon_{\mathbf{G}}\varepsilon_{C^\circ_{\mathbf{G}}(s)} \operatorname{St}_{C^\circ_{\mathbf{G}}(s)}(x)$$

by 8.16. So we may assume that s is central in $\mathbf{G}$. But then there exists a rational Borel subgroup $\mathbf{B}$ which contains x. Indeed the unipotent part of x is contained in a rational Borel subgroup as is any rational unipotent element (see 3.20), and s, being central, is contained in any Borel subgroup. So by lemma 9.2 we have $\mathrm{St}_{\mathbf{G}}(x) = (\mathrm{Res}^{\mathbf{G}^F}_{\mathbf{B}^F} \mathrm{St}_{\mathbf{G}})(x) = (\mathrm{Ind}^{\mathbf{B}^F}_{\mathbf{T}^F} \mathrm{Id}_{\mathbf{T}})(x)$. Thus $\mathrm{St}_{\mathbf{G}}(x) = 0$ if x has no conjugate in $\mathbf{T}^F$, *i.e.*, is not semi-simple, and if $x = s$ we get, by 3.19,

$$\mathrm{St}_{\mathbf{G}}(x) = |\mathbf{B}^F|/|\mathbf{T}^F| = |\mathbf{G}^F|_p,$$

whence the result. ∎

9.4 COROLLARY (OF 9.3). *The dual of the regular representation* $\mathrm{reg}_{\mathbf{G}}$ *of* $\mathbf{G}^F$ *is* $D_{\mathbf{G}}(\mathrm{reg}_{\mathbf{G}}) = \gamma_p$, *where* γ_p *denotes the element of* $\mathcal{C}(\mathbf{G}^F)$ *whose value is* $|\mathbf{G}^F|_{p'}$ *on unipotent elements and* 0 *elsewhere.*

PROOF: By 7.4 and 7.5 we have $D_{\mathbf{G}}(\chi.f) = D_{\mathbf{G}}(\chi).f$ for any $\chi \in \mathcal{C}(\mathbf{G}^F)$ and any $f \in \mathcal{C}(\mathbf{G}^F)_{p'}$. So, since clearly $\gamma_p \in \mathcal{C}(\mathbf{G}^F)_{p'}$, we have

$$D_{\mathbf{G}}(\gamma_p) = D_{\mathbf{G}}(\mathrm{Id}_{\mathbf{G}}.\gamma_p) = D_{\mathbf{G}}(\mathrm{Id}_{\mathbf{G}})\gamma_p = \mathrm{St}_{\mathbf{G}}.\gamma_p = \mathrm{reg}_{\mathbf{G}},$$

the last equality by 9.3. ∎

9.5 COROLLARY (OF 9.4). *The number of unipotent elements in* $\mathbf{G}^F$ *is equal to* $(|\mathbf{G}^F|_p)^2$.

PROOF: This formula is easily deduced from 9.4 by writing

$$\langle \mathrm{reg}_{\mathbf{G}}, \mathrm{reg}_{\mathbf{G}} \rangle = \langle \gamma_p, \gamma_p \rangle.$$

∎

9.6 PROPOSITION. *For any character* χ *of* $\mathbf{L}^F$ *we have*

$$\mathrm{St}_{\mathbf{G}}.R^{\mathbf{G}}_{\mathbf{L}}(\chi) = \mathrm{Ind}^{\mathbf{G}^F}_{\mathbf{L}^F}(\mathrm{St}_{\mathbf{L}}.\chi).$$

PROOF: Let $\mathcal{C}(\mathbf{G}^F_{p'})$ be the space of class functions on $\mathbf{G}^F_{p'}$ (the set of elements of order prime to p, which is also the set of semi-simple elements of $\mathbf{G}^F$; see 3.18); we identify $\mathcal{C}(\mathbf{G}^F_{p'})$ with the subspace of $\mathcal{C}(\mathbf{G}^F)$ consisting of functions which are zero outside $\mathbf{G}^F_{p'}$. For $f \in \mathcal{C}(\mathbf{G}^F_{p'})$, let us denote by $\mathrm{Br}\, f$ the element of $\mathcal{C}(\mathbf{G}^F)$ defined by $(\mathrm{Br}\, f)(x) = f(s)$ where s is the semi-simple part of x (Br is similar to the "Brauer lifting" in block theory). Then we have $\mathrm{Br}\, f = D_{\mathbf{G}}(\mathrm{St}_{\mathbf{G}}.f)$. Indeed $\mathrm{Br}\, f \in \mathcal{C}(\mathbf{G}^F)_{p'}$, so, as in the proof of 9.4 we have $D_{\mathbf{G}}(\mathrm{Id}_{\mathbf{G}}.\mathrm{Br}\, f) = D_{\mathbf{G}}(\mathrm{Id}_{\mathbf{G}}).\mathrm{Br}\, f = \mathrm{St}_{\mathbf{G}}.\mathrm{Br}\, f = \mathrm{St}_{\mathbf{G}}.f$, whence the result as $D_{\mathbf{G}}$ is an involution. So for any $\chi \in \mathcal{C}(\mathbf{G}^F)$ we have $D_{\mathbf{G}}(\chi.\mathrm{St}_{\mathbf{G}}) = \mathrm{Br}(\chi|_{\mathbf{G}^F_{p'}})$. But for any $f \in \mathcal{C}(\mathbf{G}^F)_{p'}$, as $\mathrm{Br}\, f \in \mathcal{C}(\mathbf{G}^F)_{p'}$, we have, by 7.4,

$${}^*R^{\mathbf{G}}_{\mathbf{L}}(\mathrm{Br}\, f) = \mathrm{Res}^{\mathbf{G}^F}_{\mathbf{L}^F} \mathrm{Br}\, f = \mathrm{Br}(\mathrm{Res}^{\mathbf{G}^F_{p'}}_{\mathbf{L}^F_{p'}} f).$$

If we apply 8.11 and then the two previous formulae we find

$$D_{\mathbf{L}}({}^*R_{\mathbf{L}}^{\mathbf{G}}(\chi.\mathrm{St}_{\mathbf{G}})) = {}^*R_{\mathbf{L}}^{\mathbf{G}}(D_{\mathbf{G}}(\chi.\mathrm{St}_{\mathbf{G}})) = {}^*R_{\mathbf{L}}^{\mathbf{G}}(\mathrm{Br}(\chi|_{\mathbf{G}^F_{p'}}))$$
$$= \mathrm{Br}(\chi|_{\mathbf{L}^F_{p'}}) = D_{\mathbf{L}}(\mathrm{St}_{\mathbf{L}} \,.\, \mathrm{Res}_{\mathbf{L}^F}^{\mathbf{G}^F} \chi),$$

which gives after applying $D_{\mathbf{L}}$

$${}^*R_{\mathbf{L}}^{\mathbf{G}}(\chi.\mathrm{St}_{\mathbf{G}}) = (\mathrm{Res}_{\mathbf{L}^F}^{\mathbf{G}^F} \chi).\, \mathrm{St}_{\mathbf{L}},$$

whence the proposition by adjunction since the multiplication by $\mathrm{St}_{\mathbf{G}}$ is self-adjoint. ■

References

The Steinberg character was originally defined by Steinberg in [St1]. Its definition as the dual of the trivial character is due to Curtis. Our exposition here (in particular in the proofs of 9.3, 9.4, 9.5 and 9.6) follows the ideas of [DM1], making a systematic use of the properties of the duality functor.

10. ℓ-ADIC COHOMOLOGY

Let $\mathbf{X}$ be an algebraic variety over $\overline{\mathbb{F}}_q$ and ℓ be a prime number different from the characteristic p of $\overline{\mathbb{F}}_q$. Using the results of Grothendieck one can associate to $\mathbf{X}$ groups of ℓ-adic cohomology with compact support $H^i_c(\mathbf{X}, \overline{\mathbb{Q}}_\ell)$ (see [SGA4$\frac{1}{2}$, Rapport, 2.10]) which are $\overline{\mathbb{Q}}_\ell$-vector spaces of finite dimension. This cohomology theory is a powerful tool. It will be used in the next chapter to define Deligne-Lusztig induction. We recall in this chapter those properties of ℓ-adic cohomology that we need . Throughout this chapter $\mathbf{X}$ will stand for an algebraic variety over $\overline{\mathbb{F}}_q$. The reader who wants to keep within the framework of chapter 3 may assume that all the varieties in this chapter are quasi-projective.

10.1 Proposition. *We have* $H^i_c(\mathbf{X}, \overline{\mathbb{Q}}_\ell) = 0$ *if* $i \notin \{0, \ldots, 2\dim \mathbf{X}\}$.

Proof: See [SGA4, XVII, 5.2.8.1]. ■

We shall put $H^*_c(\mathbf{X}) = \sum_i (-1)^i H^i_c(\mathbf{X}, \overline{\mathbb{Q}}_\ell)$; it is a virtual $\overline{\mathbb{Q}}_\ell$-vector space of finite dimension. Henceforth we shall write $H^i_c(\mathbf{X})$ for $H^i_c(\mathbf{X}, \overline{\mathbb{Q}}_\ell)$, since we shall consider only cohomology with coefficients in $\overline{\mathbb{Q}}_\ell$.

10.2 Proposition. *Any finite morphism* $\mathbf{X} \to \mathbf{X}$ *induces a linear endomorphism of* $H^i_c(\mathbf{X})$ *for any* i, *and this correspondence is functorial; a Frobenius endomorphism induces an automorphism of this space.*

Proof: The first assertion expresses the functoriality of ℓ-adic cohomology. For the second assertion see, *e.g.*, [SGA4$\frac{1}{2}$, Rapport, 1.2]. ■

10.3 Definition. *If* $g \in \mathrm{Aut}(\mathbf{X})$ *is of finite order, we call the* **Lefschetz number** *of* g *the number* $\mathcal{L}(g, \mathbf{X}) = \mathrm{Trace}(g \mid H^*_c(\mathbf{X}))$.

The fundamental property of ℓ-adic cohomology is the "Lefschetz theorem":

10.4 Theorem. *Let* F *be the Frobenius endomorphism associated to an* $\mathbb{F}_q$*-structure on* $\mathbf{X}$. *We have* $|\mathbf{X}^F| = \mathrm{Trace}(F \mid H^*_c(\mathbf{X}))$.

Proof: See [SGA4$\frac{1}{2}$, Rapport, 3.2]. ■

10.5 Corollary. *Assume that* $\mathbf{X}$ *is defined over* $\mathbb{F}_q$, *with Frobenius endomorphism* F, *and that* $g \in \mathrm{Aut}(\mathbf{X})$ *is a rational automorphism of finite* *order; then we have $\mathcal{L}(g,\mathbf{X}) = R(t)|_{t=\infty}$, where $R(t)$ is the formal series $-\sum_{n=1}^{\infty}|\mathbf{X}^{gF^n}|t^n$.*

This result is an example of the power of ℓ-adic cohomology: it proves that $R(t)|_{t=\infty}$ is the value at g of some character for any finite subgroup of $\mathrm{Aut}(\mathbf{X})$ containing g. No other proof of this fact is known.

PROOF: As g and F commute, they can be reduced to a triangular form in the same basis of $\oplus_i H_c^i(\mathbf{X})$ (note that F is not necessarily semi-simple). Let $\lambda_1,\ldots,\lambda_k$ be the eigenvalues of F and $x_1,\ldots,x_k$ be those of g, and let $\varepsilon_i = \pm 1$ be the sign of the cohomology space (in $H_c^*(\mathbf{X})$) in which λ_i and x_i are eigenvalues. As gF^n is also a Frobenius endomorphism on $\mathbf{X}$ for any n (see 3.6 (i)), we have by 10.4

$$R(t) = -\sum_{n=1}^{\infty}\sum_{i=1}^{k}\varepsilon_i\lambda_i^n x_i t^n = \sum_{i=1}^{k}\varepsilon_i x_i \frac{-\lambda_i t}{1-\lambda_i t}$$

whence $R(t)|_{t=\infty} = \sum_{i=1}^{k}\varepsilon_i x_i$. ∎

10.6 COROLLARY. *The Lefschetz number $\mathcal{L}(g,\mathbf{X})$ is a rational integer independent of ℓ.*

PROOF: The independence of ℓ is a clear consequence of 10.5 as, given g, there always exists a rational structure on $\mathbf{X}$ over some finite subfield of $\overline{\mathbb{F}}_q$ such that g is rational, see 3.6 (iv). Moreover the proof of 10.5 shows that $R(t)$ is in $\overline{\mathbb{Q}}_\ell(t)$. As it is a formal series with integral coefficients, it has to be in $\mathbb{Q}(t)$. So we have $\mathcal{L}(g,\mathbf{X}) \in \mathbb{Q}$. But a Lefschetz number is an algebraic integer since it is equal to $\sum_{i=1}^{k}\varepsilon_i x_i$ where all x_i are roots of unity (of the same order as g), whence the result. ∎

In the following propositions we shall list properties of the Lefschetz numbers. When these properties can be proved directly using 10.5, we shall give that proof; but we shall also give, without proof, the corresponding result on ℓ-adic cohomology, if it exists.

10.7 PROPOSITION.

(i) *Let $\mathbf{X} = \mathbf{X}_1 \coprod \mathbf{X}_2$ be a partition of $\mathbf{X}$ into two subvarieties with $\mathbf{X}_1$ open (so $\mathbf{X}_2$ closed). We have a long exact sequence*

$$\ldots \to H_c^i(\mathbf{X}_1) \to H_c^i(\mathbf{X}) \to H_c^i(\mathbf{X}_2) \to H_c^{i+1}(\mathbf{X}_1) \to \ldots,$$

whence $H_c^*(\mathbf{X}) = H_c^*(\mathbf{X}_1) + H_c^*(\mathbf{X}_2)$ as virtual vector spaces.

(ii) *If in (i) $\mathbf{X}_1$ is also closed, then the exact sequence of (i) splits, and for any i we have $H_c^i(\mathbf{X}) \simeq H_c^i(\mathbf{X}_1) \oplus H_c^i(\mathbf{X}_2)$.*

(iii) *Let* $\mathbf{X} = \coprod_j \mathbf{X}_j$ *be a finite partition of* $\mathbf{X}$ *into locally closed subvarieties; if* g *is an automorphism of finite order of* $\mathbf{X}$ *which stabilizes (globally) this partition, we have* $\mathcal{L}(g, \mathbf{X}) = \sum_{\{j \mid {}^g\mathbf{X}_j = \mathbf{X}_j\}} \mathcal{L}(g, \mathbf{X}_j)$.

PROOF: For (i) and (ii) see, *e.g.*, [SGA4, XVII, 5.1.16.3]. We prove (iii). Let F be a Frobenius endomorphism which commutes with g and is such that all $\mathbf{X}_j$ are rational; see 3.6 (iv). Assertion (iii) is clear from 10.5 and the equality $|\mathbf{X}^{gF^n}| = \sum_{\{j \mid {}^g\mathbf{X}_j = \mathbf{X}_j\}} |\mathbf{X}_j^{gF^n}|$. ■

10.8 PROPOSITION. *Assume that the variety* $\mathbf{X}$ *is reduced to a finite number of points.*

(i) *We have* $H^i_c(\mathbf{X}) = 0$ *if* $i \neq 0$ *and* $H^0_c(\mathbf{X}) \simeq \overline{\mathbb{Q}}_\ell[\mathbf{X}]$.

(ii) *Any permutation* g *of the finite set* $\mathbf{X}$ *defines an automorphism of the variety* $\mathbf{X}$*, and* $H^*_c(\mathbf{X}) \simeq \overline{\mathbb{Q}}_\ell[\mathbf{X}]$ *is a permutation module under* g*. We have* $\mathcal{L}(g, \mathbf{X}) = |\mathbf{X}^g|$.

PROOF: These results are clear from 10.1 and 10.7. ■

10.9 PROPOSITION. *Let* $\mathbf{X}$ *and* $\mathbf{X}'$ *be two varieties.*

(i) *We have* $H^k_c(\mathbf{X} \times \mathbf{X}') \simeq \bigoplus_{i+j=k} H^i_c(\mathbf{X}) \otimes_{\overline{\mathbb{Q}}_\ell} H^j_c(\mathbf{X}')$ *(the "Künneth formula").*

(ii) *Let* $g \in \operatorname{Aut}\mathbf{X}$ *(resp.* $g' \in \operatorname{Aut}\mathbf{X}'$*) be automorphisms of finite order; then we have* $\mathcal{L}(g \times g', \mathbf{X} \times \mathbf{X}') = \mathcal{L}(g, \mathbf{X})\mathcal{L}(g', \mathbf{X}')$.

PROOF: For (i) see [SGA4, XVII, 5.4.3]. Let us prove (ii). We write $f * h = \sum_{i \geq 0} a_i b_i t^i$ for the Hadamard product of two formal series $f = \sum_{i \geq 0} a_i t^i$ and $h = \sum_{i \geq 0} b_i t^i$. We have to show that the series $f = \sum_{n \geq 1} |\mathbf{X}^{gF^n}| t^n$ and $h = \sum_{n \geq 1} |\mathbf{X}'^{g'F^n}| t^n$ satisfy the relation $-(f * h)|_{t=\infty} = -f|_{t=\infty} \times -h|_{t=\infty}$. But this result follows easily from the proof of 10.5, because these two series are linear combinations of series of the form $t/(1-\lambda t)$ and such series satisfy the equality. ■

10.10 PROPOSITION. *Let* $H \subset \operatorname{Aut}\mathbf{X}$ *be a finite subgroup such that the quotient variety* $\mathbf{X}/H$ *exists (which is always true if* $\mathbf{X}$ *is quasi-projective). Consider an automorphism* $g \in \operatorname{Aut}\mathbf{X}$ *of finite order which commutes with all elements of* H*. Then:*

(i) *The* $\overline{\mathbb{Q}}_\ell[g]$*-modules* $H^i_c(\mathbf{X})^H$ *and* $H^i_c(\mathbf{X}/H)$ *are isomorphic for any* i,

(ii) $\mathcal{L}(g, \mathbf{X}/H) = |H|^{-1} \sum_{h \in H} \mathcal{L}(gh, \mathbf{X})$.

PROOF: For (i) see, *e.g.*, [SGA4, XVII, 6.2.5]. Let us prove (ii). Let us choose a rational structure on $\mathbf{X}$ such that g and all elements of H are rational. If F is the associated Frobenius endomorphism, it induces on

$\mathbf{X}/H$ a rational structure (*e.g.*, by 3.3 (i)), and we have

$$|(\mathbf{X}/H)^{gF^n}| = |H|^{-1} \sum_{h\in H} |\mathbf{X}^{ghF^n}|,$$

whence the result. ■

10.11 PROPOSITION. *Let* $\mathbf{X}$ *be an affine space of dimension* n*; then:*

(i) $\dim H^i_c(\mathbf{X}) = \begin{cases} 1, & \text{if } i = 2n, \\ 0, & \text{otherwise.} \end{cases}$

(ii) *If* F *is the Frobenius endomorphism associated to some* $\mathbb{F}_q$*-structure on* $\mathbf{X}$*, we have* $|\mathbf{X}^F| = q^n$*.*

(iii) *For any finite order* $g \in \mathrm{Aut}(\mathbf{X})$ *we have* $\mathcal{L}(g, \mathbf{X}) = 1$*.*

PROOF: For (i) see, *e.g.*, [Sr, 5.7]. Note that (iii) is straightforward from (ii). Let us prove (ii). Let λ be the scalar by which F acts on the one-dimensional space $H^{2n}_c(\mathbf{X})$; for any $m > 0$ we have $|\mathbf{X}^{F^m}| = \lambda^m$ so λ is a positive integer. We now show that, if $A = A_\circ \otimes_{\mathbb{F}_q} \overline{\mathbb{F}}_q$ defines the $\mathbb{F}_q$-structure on $\mathbf{X}$, there exists n_0 such that $A_0 \otimes_{\mathbb{F}_q} \mathbb{F}_{q^{n_0}} \simeq \mathbb{F}_{q^{n_0}}[T_1, \ldots, T_n]$. Indeed we have $A \simeq \overline{\mathbb{F}}_q[T_1, \ldots, T_n]$, so if $A_\circ$ is generated by $t_1, \ldots, t_k$ and if n_0 is such that all t_i are in $\mathbb{F}_{q^{n_0}}[T_1, \ldots, T_n]$ we have $A_\circ \otimes_{\mathbb{F}_q} \mathbb{F}_{q^{n_0}} \subset \mathbb{F}_{q^{n_0}}[T_1, \ldots, T_n]$ and this inclusion has to be an equality because these two $\mathbb{F}_{q^{n_0}}$-spaces have equal tensor products with $\overline{\mathbb{F}}_q$. For any m multiple of n_0 we have $|\mathbf{X}^{F^m}| = q^{mn}$, so $\lambda = q^n$, which proves (ii) by 10.4. ■

10.12 PROPOSITION. *Let* $\mathbf{X} \xrightarrow{\pi} \mathbf{X}'$ *be an epimorphism such that all fibres are isomorphic to affine spaces of the same dimension* n*. Let* $g \in \mathrm{Aut}\,\mathbf{X}$ *(resp.* $g' \in \mathrm{Aut}\,\mathbf{X}'$*) be finite order automorphisms such that* $g'\pi = \pi g$*.*

(i) *The* $\overline{\mathbb{Q}}_\ell[g]$*-module* $H^i_c(\mathbf{X})$ *is isomorphic to* $H^{i-2n}_c(\mathbf{X}')(-n)$*, a "Tate twist" of the module* $H^j_c(\mathbf{X}')$ *(if* g *acts on this last module by the action of* g'*).* (See [Sr, III page 47] for the definition of a "Tate twist"; note that $H^j_c(\mathbf{X}')(-n)$ and $H^j_c(\mathbf{X}')$ are isomorphic vector spaces and that for any $\mathbb{F}_q$-structure on $\mathbf{X}'$ the action of F on $H^j_c(\mathbf{X}')(-n)$ is q^n times the action of F on $H^j_c(\mathbf{X}')$.)

(ii) *We have* $\mathcal{L}(g, \mathbf{X}) = \mathcal{L}(g', \mathbf{X}')$*.*

PROOF: For (i) see [Sr, 5.5, 5.7]. Let us prove (ii). We choose rational structures on $\mathbf{X}$ and $\mathbf{X}'$ over the same field $\mathbb{F}_q$ such that π is rational. By 10.11 (ii) we have

$$|\mathbf{X}^{gF^m}| = \sum_{y \in \mathbf{X}'^{g'F^m}} |\pi^{-1}(y)^{gF^m}| = |\mathbf{X}'^{g'F^m}| q^{mn},$$

whence the result. ■

10.13 PROPOSITION. *Let* $\mathbf{G}$ *be a connected algebraic group acting on the variety* $\mathbf{X}$.

(i) *Any element* $g \in \mathbf{G}$ *acts trivially on* $H^i_c(\mathbf{X})$ *for any* i.

(ii) *Let* $g \in \mathbf{G}$ *be such that the induced isomorphism is of finite order; then* $\mathcal{L}(g, \mathbf{X}) = \mathcal{L}(1, \mathbf{X})$.

PROOF: The proof of (i) may be found in [DL1, 6.4, 6.5]. Let us prove (ii). We choose rational structures on $\mathbf{G}$ and $\mathbf{X}$ such that the action of $\mathbf{G}$ is rational (*e.g.*, apply 3.6 (iv)), *i.e.*, we have ${}^F(gx) = {}^Fg{}^Fx$ for all $(g, x) \in \mathbf{G} \times \mathbf{X}$. If n is a positive integer, by the Lang-Steinberg theorem 3.10 there exists $h \in \mathbf{G}$ such that $h.{}^{F^n}h^{-1} = g$. But then $x \mapsto h^{-1}x$ defines a bijection from $\mathbf{X}^{gF^n}$ onto $\mathbf{X}^{F^n}$, so we have $|\mathbf{X}^{gF^n}| = |\mathbf{X}^{F^n}|$, whence the result. ∎

10.14 PROPOSITION. *Let* $g = su$ *be the decomposition of the finite order automorphism* $g \in \operatorname{Aut} \mathbf{X}$ *into the product of a* p'*-element* s *and a* p*-element* u. *Then we have* $\mathcal{L}(g, \mathbf{X}) = \mathcal{L}(u, \mathbf{X}^s)$.

PROOF: This result cannot be proved by means of a computation using only Lefschetz numbers. One has to use directly the definition of ℓ-adic cohomology (see [DL1, 3.1]). ∎

10.15 PROPOSITION. *Let* $\mathbf{T}$ *be a torus acting on a variety* $\mathbf{X}$ *and let* $g \in \operatorname{Aut} \mathbf{X}$ *be a finite order automorphism commuting with the action of* $\mathbf{T}$. *Then for any* $t \in \mathbf{T}$ *we have* $\mathcal{L}(g, \mathbf{X}) = \mathcal{L}(g, \mathbf{X}^t)$. *Moreover, if* $\mathbf{X}$ *is affine, we have* $\mathcal{L}(g, \mathbf{X}) = \mathcal{L}(g, \mathbf{X}^{\mathbf{T}})$.

PROOF: Let $g = su$ be the decomposition of g into its p'-part and its p-part; by 10.14 we have $\mathcal{L}(g, \mathbf{X}) = \mathcal{L}(u, \mathbf{X}^s)$. The action of $\mathbf{T}$ commutes with g, so commutes also with s, thus $\mathbf{T}$ acts on $\mathbf{X}^s$. As this group is connected the action of any $t \in \mathbf{T}$ on $H^*_c(\mathbf{X}^s)$ is trivial by 10.13. Hence for any $t \in \mathbf{T}$ we have

$$\mathcal{L}(u, \mathbf{X}^s) = \mathcal{L}(ut, \mathbf{X}^s) = \mathcal{L}(u, (\mathbf{X}^s)^t) = \mathcal{L}(u, (\mathbf{X}^t)^s) = \mathcal{L}(g, \mathbf{X}^t),$$

the second and the last equalities coming from 10.14. If $\mathbf{X}$ is affine 0.7 shows that there exists $t \in \mathbf{T}$ such that $\mathbf{X}^{\mathbf{T}} = \mathbf{X}^t$. ∎

References

The theory of ℓ-adic cohomology is expounded in [SGA4], [SGA4$\frac{1}{2}$] and [SGA5]. For the properties of the Lefschetz numbers one can also refer to [L2]. A good survey, with some proofs, of properties needed in our case may be found in [Sr, chapter V].

11. DELIGNE-LUSZTIG INDUCTION; THE MACKEY FORMULA

We are going now to extend the construction of the functor $R_{\mathbf{L}}^{\mathbf{G}}$ to the case where $\mathbf{L}$ is a rational Levi subgroup of $\mathbf{G}$ which is not the Levi subgroup of any rational parabolic subgroup of $\mathbf{G}$.

EXAMPLE. We would like to generalize the construction of 6.2 to the case of the unitary groups (see 15.1), but the construction used there does not work, since the Frobenius morphism on $\mathbf{U}_{n+m}$ maps the parabolic subgroup $\begin{pmatrix} \mathbf{U}_n & * \\ 0 & \mathbf{U}_m \end{pmatrix}$ to the parabolic subgroup $\begin{pmatrix} \mathbf{U}_n & 0 \\ * & \mathbf{U}_m \end{pmatrix}$ which is not conjugate to it if $n \neq m$. Actually, in this case the rational Levi subgroup $\begin{pmatrix} \mathbf{U}_n & 0 \\ 0 & \mathbf{U}_m \end{pmatrix}$ is not a Levi subgroup of any rational parabolic subgroup.

The idea that Deligne and Lusztig had was to associate to any parabolic subgroup $\mathbf{P}$ with $\mathbf{L}$ as a Levi subgroup an algebraic variety $\mathbf{X}$ on $\overline{\mathbb{F}}_q$, endowed with an action of $\mathbf{G}^F \times \mathbf{L}^{F\text{opp}}$ such that when $\mathbf{P} = {}^F\mathbf{P}$ we have $H_c^*(\mathbf{X}) \simeq \overline{\mathbb{Q}}_\ell[\mathbf{G}^F/\mathbf{U}^F]$, and to define $R_{\mathbf{L}}^{\mathbf{G}}$ as the functor associated to the $\mathbf{G}^F$-module-$\mathbf{L}^F$ afforded by $H_c^*(\mathbf{X})$. This construction of Deligne and Lusztig was first published by Lusztig (see [L1]) and is traditionally called the "Lusztig functor" (or the "Lusztig induction"). More precisely:

11.1 DEFINITION. *The Lusztig functor $R_{\mathbf{L}\subset\mathbf{P}}^{\mathbf{G}}$, where $\mathbf{P} = \mathbf{LU}$ is a Levi decomposition of $\mathbf{P}$, is the generalized induction functor associated to the $\mathbf{G}^F$-module-$\mathbf{L}^F$ afforded by $H_c^*(\mathcal{L}^{-1}(\mathbf{U}))$, where $\mathcal{L} : \mathbf{G} \to \mathbf{G}$ is the Lang map $x \mapsto x^{-1}\,{}^F x$, and where the action of $(g, l) \in \mathbf{G}^F \times \mathbf{L}^{F\text{opp}}$ is induced by that on $\mathcal{L}^{-1}(\mathbf{U})$ given by $x \mapsto gxl$.*

The adjoint functor is denoted by ${}^*R_{\mathbf{L}\subset\mathbf{P}}^{\mathbf{G}}$ (and sometimes called the "Lusztig restriction").

We note that if $\mathbf{P}$ is rational then $\mathbf{U}$ is also, in which case if $x \in \mathcal{L}^{-1}(\mathbf{U})$ then xu is also for any $u \in \mathbf{U}$. We thus get a map $x \mapsto x\mathbf{U} : \mathcal{L}^{-1}(\mathbf{U}) \to \mathbf{G}/\mathbf{U}$, whose image is $(\mathbf{G}/\mathbf{U})^F \simeq \mathbf{G}^F/\mathbf{U}^F$ and whose fibres are all isomorphic to the affine space $\mathbf{U}$ (see 0.31 (i) and 0.33). Applying 10.12 (i) and 10.8 we get $H_c^*(\mathbf{X}) \simeq \overline{\mathbb{Q}}_\ell[\mathbf{G}^F/\mathbf{U}^F]$ as $\mathbf{G}^F$-modules-$\mathbf{L}^F$, which proves that the Lusztig induction is indeed a generalization of the Harish-Chandra induction.

From 4.5 we immediately obtain the following formulae.

11.2 PROPOSITION. *With the above notation, we have*

$$(R^{\mathbf{G}}_{\mathbf{L}\subset\mathbf{P}}\chi)(g) = |\mathbf{L}^F|^{-1} \sum_{l\in\mathbf{L}^F} \operatorname{Trace}((g,l) \mid H^*_c(\mathcal{L}^{-1}(\mathbf{U})))\chi(l^{-1})$$

and

$$({}^*R^{\mathbf{G}}_{\mathbf{L}\subset\mathbf{P}}\psi)(l) = |\mathbf{G}^F|^{-1} \sum_{g\in\mathbf{G}^F} \operatorname{Trace}((g,l) \mid H^*_c(\mathcal{L}^{-1}(\mathbf{U})))\psi(g^{-1}).$$

We will need the fact that the $\mathbf{L}^F$-module-$\mathbf{G}^F$ affording ${}^*R^{\mathbf{G}}_{\mathbf{L}\subset\mathbf{P}}$ can also be defined with the "dual variety".

11.3 PROPOSITION. *The $\mathbf{L}^F$-module-$\mathbf{G}^F$ afforded by $H^*_c(\mathcal{L}^{-1}(\mathbf{U}))$ for the action induced from that on $\mathcal{L}^{-1}(\mathbf{U})$ given, for $(g,l) \in \mathbf{G}^{F\mathrm{opp}} \times \mathbf{L}^F$, by $x \mapsto g^{-1}xl^{-1}$, is isomorphic in the Grothendieck group of $\mathbf{L}^F$-modules-$\mathbf{G}^F$ to the module which defines ${}^*R^{\mathbf{G}}_{\mathbf{L}\subset\mathbf{P}}$.*

PROOF: As Lefschetz numbers are integers, (g,l) and (g^{-1},l^{-1}) have the same trace on the cohomology of $\mathcal{L}^{-1}(\mathbf{U})$. But the trace of an element is the same on a given module as on its dual. Thus the module dual to that defining $R^{\mathbf{G}}_{\mathbf{L}\subset\mathbf{P}}$ and the module of the statement afford the same character. They are thus isomorphic, since group algebras over an algebraically closed field of characteristic 0 are semi-simple so an element of the Grothendieck group is defined by its character. ■

We will denote by $\mathcal{L}^{-1}(\mathbf{U})^\vee$ the variety $\mathcal{L}^{-1}(\mathbf{U})$ endowed with the above action of $\mathbf{G}^F \times \mathbf{L}^F$.

The next proposition also results from the fact that the Lefschetz numbers are integers.

11.4 PROPOSITION. *Let $\pi \in \operatorname{Irr}(\mathbf{L}^F)$ and let $\overline{\pi}$ be the contragredient representation; then the contragredient of $R^{\mathbf{G}}_{\mathbf{L}\subset\mathbf{P}}(\pi)$ is $R^{\mathbf{G}}_{\mathbf{L}\subset\mathbf{P}}(\overline{\pi})$.*

PROOF: The proposition results immediately from formula 11.2 and from the fact that the trace of an element x on a given representation is equal to the trace of x^{-1} on the contragredient representation. ■

11.5 TRANSITIVITY. *Let $\mathbf{Q} \subset \mathbf{P}$ be two parabolic subgroups of $\mathbf{G}$, and let $\mathbf{M}$ (resp. $\mathbf{L}$) be a rational Levi subgroup of $\mathbf{Q}$ (resp. $\mathbf{P}$). Assume that $\mathbf{M} \subset \mathbf{L}$. Then $R^{\mathbf{G}}_{\mathbf{L}\subset\mathbf{P}} \circ R^{\mathbf{L}}_{\mathbf{M}\subset\mathbf{L}\cap\mathbf{Q}} = R^{\mathbf{G}}_{\mathbf{M}\subset\mathbf{Q}}$.*

PROOF: According to 4.4 (see 4.7) we must show that there is an isomorphism of $\mathbf{G}^F$-modules-$\mathbf{M}^F$

$$H^*_c(\mathcal{L}^{-1}(\mathbf{U})) \otimes_{\overline{\mathbb{Q}}_\ell[\mathbf{L}^F]} H^*_c(\mathcal{L}^{-1}(\mathbf{V}\cap\mathbf{L})) \simeq H^*_c(\mathcal{L}^{-1}(\mathbf{V})),$$

where $\mathbf{P} = \mathbf{LU}$ (resp. $\mathbf{Q} = \mathbf{MV}$) is the Levi decomposition of $\mathbf{P}$ (resp. $\mathbf{Q}$). This results from the isomorphism of "$\mathbf{L}^F$-varieties-$\mathbf{M}^F$" given by

$$\mathcal{L}_{\mathbf{G}}^{-1}(\mathbf{U}) \times_{\mathbf{L}^F} \mathcal{L}_{\mathbf{L}}^{-1}(\mathbf{V} \cap \mathbf{L}) \simeq \mathcal{L}^{-1}(\mathbf{V})$$
$$(x, l) \mapsto xl,$$

(where the notation is as in 0.42 and where $\mathcal{L}^{-1}$ has been indexed to specify in which group we take the inverse image) when we apply the properties 10.9 and 10.10 of the cohomology. (We may apply 10.10 since $\mathcal{L}^{-1}(\mathbf{U})$ and $\mathcal{L}^{-1}(\mathbf{V} \cap \mathbf{L})$ are affine as closed subvarieties of $\mathbf{G}$.) ■

11.6. The rest of this chapter is devoted to the study of the "Mackey formula" (5.1) for the Lusztig functors. At the present time, we do not know of a proof in all cases; we are going to give those reductions that we can make and deduce the formula in some known cases. It is conjectured that the formula always holds; in addition to the results proved below, there is an "asymptotic" proof (for p and q large enough) due to Deligne, and verifications in some special cases (*e.g.*, $\mathbf{GL}_n$ and $\mathbf{U}_n$ in [DM2, 2.1]).

Let us first consider the left-hand side, *i.e.*, the composite functor ${}^*R^{\mathbf{G}}_{\mathbf{L}\subset\mathbf{P}} \circ R^{\mathbf{G}}_{\mathbf{L}'\subset\mathbf{P}'}$ where $\mathbf{P} = \mathbf{LU}$ and $\mathbf{P}' = \mathbf{L}'\mathbf{U}'$ are two Levi decompositions with $\mathbf{L}$ and $\mathbf{L}'$ rational. Then, by 4.4, this composite functor is associated to the $\mathbf{L}^F$-module-$\mathbf{L}'^F$ given by $H_c^*((\mathcal{L}^{-1}(\mathbf{U}))^\vee) \otimes_{\overline{\mathbb{Q}}_\ell[\mathbf{G}^F]} H_c^*(\mathcal{L}^{-1}(\mathbf{U}'))$ where $(g, l') \in \mathbf{G}^F \times \mathbf{L}'^{F\text{opp}}$ acts on $\mathcal{L}^{-1}(\mathbf{U}')$ by $x' \mapsto gx'l'$ and where $(\mathcal{L}^{-1}(\mathbf{U}))^\vee$ is as defined after 11.3. This tensor product of cohomology spaces is isomorphic, by the Künneth formula 10.9 and by 10.10 (see proof of 11.5 above), to the $\mathbf{L}^F$-module-$\mathbf{L}'^F$ given by $H_c^*(\mathcal{L}^{-1}(\mathbf{U}) \times_{\mathbf{G}^F} \mathcal{L}^{-1}(\mathbf{U}'))$ (the quotient by the action of $g \in \mathbf{G}^F$ on $\mathcal{L}^{-1}(\mathbf{U}) \times \mathcal{L}^{-1}(\mathbf{U}')$ identifies (x, x') with (gx, gx')).

11.7 LEMMA. *Let $\mathbf{Z} = \{(u, u', g) \in \mathbf{U} \times \mathbf{U}' \times \mathbf{G} \mid u.{}^Fg = gu'\}$. The map $\varphi : (x, x') \mapsto (\mathcal{L}(x), \mathcal{L}(x'), x^{-1}x')$ is an isomorphism of $\mathbf{L}^F$-varieties-$\mathbf{L}'^F$ from $\mathcal{L}^{-1}(\mathbf{U}) \times_{\mathbf{G}^F} \mathcal{L}^{-1}(\mathbf{U}')$ to $\mathbf{Z}$; it maps the action of $(l, l') \in \mathbf{L}^F \times \mathbf{L}'^F$ to the action on $\mathbf{Z}$ given by $(u, u', g) \mapsto ({}^lu, {}^{l'^{-1}}u', lgl')$.*

PROOF: Let us first consider the map on $\mathcal{L}^{-1}(\mathbf{U}) \times \mathcal{L}^{-1}(\mathbf{U}')$ defined by the same formula. The image of this map is clearly in $\mathbf{Z}$. On the other hand, an easy computation shows that (x, x') and (y, y') have the same image if and only if there is some $\gamma \in \mathbf{G}^F$ such that $(x, x') = (\gamma x, \gamma x')$, so the map factors through φ, and φ is injective. It remains to show that φ is surjective; let $(u, u', g) \in \mathbf{Z}$ and let x and x' be any elements of $\mathbf{G}$ such that $u = \mathcal{L}(x)$ and $u' = \mathcal{L}(x')$. We also have $u = \mathcal{L}(\gamma x)$ for any $\gamma \in \mathbf{G}^F$, so it is enough to show that we may choose γ such that $x^{-1}\gamma^{-1}x' = g$; but indeed $xgx'^{-1} \in \mathbf{G}^F$ because $u.{}^Fg = gu'$. The formula for the action on $\mathbf{Z}$ is clear. ■

Since by 5.6 we have $\mathbf{G} = \coprod_{w \in \mathbf{L}\backslash\mathcal{S}(\mathbf{L},\mathbf{L}')/\mathbf{L}'} {}^{F^{-1}}\mathbf{P} w^{F^{-1}}\mathbf{P}'$, we have

$$\mathbf{Z} = \coprod_{w \in \mathbf{L}\backslash\mathcal{S}(\mathbf{L},\mathbf{L}')/\mathbf{L}'} \mathbf{Z}_w,$$

a union of the locally closed subvarieties

$$\mathbf{Z}_w = \{(u,u',g) \in \mathbf{U} \times \mathbf{U}' \times {}^{F^{-1}}\mathbf{P} w^{F^{-1}}\mathbf{P}' \mid u.{}^F g = gu'\}.$$

Introducing the new variables $w' = w.{}^F w^{-1}$, $F' = w'F$ and $(u_1, u'_1, g_1) = (u, {}^w u', gw^{-1})$, we have $\mathbf{Z}_w \simeq \mathbf{Z}'_w$, where

$$\mathbf{Z}'_w = \{(u_1, u'_1, g_1) \in \mathbf{U} \times {}^w\mathbf{U}' \times {}^{F^{-1}}\mathbf{P}^{F'^{-1}}({}^w\mathbf{P}') \mid u_1.{}^F g_1 = g_1 u'_1 w'\}$$

and we have used the fact that ${}^{wF^{-1}}\mathbf{P}' = {}^{F'^{-1}}({}^w\mathbf{P}')$. The action of $\mathbf{L}^F \times \mathbf{L}'^F$ on $\mathbf{Z}$ induces an action of $\mathbf{L}^F \times ({}^w\mathbf{L}')^{F'}$ on $\mathbf{Z}'_w$ which is given in the new coordinates by the same formula as before.

At this stage we may "forget w", *i.e.*, express everything in terms of the new variables $\mathbf{V} = {}^w\mathbf{U}'$, $\mathbf{Q} = {}^w\mathbf{P}'$ and $\mathbf{M} = {}^w\mathbf{L}'$. With these notations the variety

$$\mathbf{Z}'_w = \{\, (u_1, v_1, g_1) \in \mathbf{U} \times \mathbf{V} \times {}^{F^{-1}}\mathbf{P}^{F'^{-1}}\mathbf{Q} \mid u_1.{}^F g_1 = g_1 v_1 w' \,\}$$

is endowed with an action of $(l,m) \in \mathbf{L}^F \times \mathbf{M}^{F'}$ given by $(u_1, v_1, g_1) \mapsto ({}^l u_1, {}^{m^{-1}} v_1, l g_1 m)$.

We are now going to use the decomposition ${}^{F^{-1}}\mathbf{P}^{F'^{-1}}\mathbf{Q} = {}^{F^{-1}}\mathbf{U}\mathbf{L}\mathbf{M}^{F'^{-1}}\mathbf{V}$.

11.8 LEMMA. *The cohomology of the variety*

$$\mathbf{Z}''_w = \{\, (u, v, u', v', n) \in \mathbf{U} \times \mathbf{V} \times {}^{F^{-1}}\mathbf{U} \times {}^{F'^{-1}}\mathbf{V} \times \mathbf{L}\mathbf{M} \mid u.{}^F n = u' n v' v w' \,\}$$

is isomorphic as an $\mathbf{L}^F$-module-$\mathbf{M}^{F'}$ to that of $\mathbf{Z}'_w$; the isomorphism of cohomology spaces is induced by the fibration

$$\pi : \mathbf{Z}''_w \to \mathbf{Z}'_w : (u, v, u', v', n) \mapsto (u.{}^F u'^{-1}, v.{}^{F'} v', u' n v'),$$

and the isomorphism of $\mathbf{L}^F$-modules-$\mathbf{M}^{F'}$ is for the action induced by the action of $(l,m) \in \mathbf{L}^F \times \mathbf{M}^{F'}$ on $\mathbf{Z}''_w$ given by

$$(u, v, u', v', n) \mapsto ({}^l u, {}^{m^{-1}} v, {}^l u', {}^{m^{-1}} v', lnm).$$

PROOF: Once we get the isomorphism of cohomology spaces, everything else is a straightforward computation. To get that isomorphism we will use 10.12; for that we need to compute the fibres of π, *i.e.*, determine the quintuples (u,v,u',v',n) which are mapped by π to (u_1,v_1,g_1). Since, once u_1 and v_1 are given, u and v are determined by u' and v', it is equivalent to find the triples (u',v',n) such that $g_1=u'nv'$, which is in turn equivalent to finding the pairs (u',v') such that $u'^{-1}g_1v'^{-1}\in\mathbf{LM}$. Let us choose some decomposition $g_1=u_1'lmv_1'$ where $u_1'\in{}^{F^{-1}}\mathbf{U}$, $v_1'\in{}^{F'^{-1}}\mathbf{V}$, $l\in\mathbf{L}$ and $m\in\mathbf{M}$; the condition on (u',v') can be written as $l^{l^{-1}}(u'^{-1}u_1')^m(v_1'v'^{-1})m\in\mathbf{LM}$; introducing the new variables $u''={}^{l^{-1}}(u'^{-1}u_1')$ and $v''={}^m(v_1'v'^{-1})$ the condition on the pair $(u'',v'')\in{}^{F^{-1}}\mathbf{U}\times{}^{F'^{-1}}\mathbf{V}$ becomes $u''v''\in\mathbf{LM}$, *i.e.*, $u''^{-1}\mathbf{L}\cap v''\mathbf{M}\neq\emptyset$. Using 5.8 for the parabolic subgroups ${}^{F^{-1}}\mathbf{P}$ and ${}^{F'^{-1}}\mathbf{Q}$ it is easy to see that the solutions are all pairs of the form $({}^yx^{-1}a^{-1},ay)$ where $x\in{}^{F^{-1}}\mathbf{U}\cap\mathbf{M}$, $y\in{}^{F'^{-1}}\mathbf{V}\cap\mathbf{L}$ and $a\in{}^{F^{-1}}\mathbf{U}\cap{}^{F'^{-1}}\mathbf{V}$. Thus all fibres of π are isomorphic to the affine space $({}^{F^{-1}}\mathbf{U}\cap\mathbf{M})\times({}^{F'^{-1}}\mathbf{V}\cap\mathbf{L})\times({}^{F^{-1}}\mathbf{U}\cap{}^{F'^{-1}}\mathbf{V})$, whence the lemma. ∎

We do not know in general how to transform the left-hand side of the Mackey formula much further towards our goal; the idea that we will use now is to find an action on $\mathbf{Z}''_w$ of a subgroup of $\mathbf{L}\times\mathbf{M}$ whose identity component is a torus which commutes to $\mathbf{L}^F\times\mathbf{M}^{F'}$; using 10.15 we may then replace $\mathbf{Z}''_w$ by its fixed points under that torus. This will work if we can find a large enough torus. Note that the image of some quintuple $(u,v,u',v',n)\in\mathbf{Z}''_w$ under $(l,m)\in\mathbf{L}\times\mathbf{M}$ is in $\mathbf{Z}''_w$ if and only if ${}^{Fn^{-1}}(l^{-1}.{}^Fl)={}^{w'^{-1}}(m^{F'}m^{-1})$.

Let $Z(\mathbf{L})$ (resp. $Z(\mathbf{M})$) be the centre of $\mathbf{L}$ (resp. of $\mathbf{M}$) and put

$$\mathbf{H}_w=\{\,(l,m)\in Z(\mathbf{L})\times Z(\mathbf{M})\mid l^{-1}.{}^Fl={}^{w'^{-1}}(m^{F'}m^{-1})\,\}.$$

Then the identity component $\mathbf{H}^\circ_w$ is a torus (a connected subgroup of $Z(\mathbf{L})^\circ\times Z(\mathbf{M})^\circ$), and the image of $(u,v,u',v',n)\in\mathbf{Z}''_w$ under $(l,m)\in\mathbf{H}_w$ is still in $\mathbf{Z}''_w$. Indeed if we write $n=\lambda\mu$ with $(\lambda,\mu)\in\mathbf{L}\times\mathbf{M}$, the condition for the image of (u,v,u',v',n) under (l,m) to be in $\mathbf{Z}''_w$ can be written ${}^{F\lambda^{-1}}(l^{-1F}l)={}^{w'^{-1F'}\mu}(m^{F'}m^{-1})$ which holds when $(l,m)\in\mathbf{H}_w$ since $l^{-1F}l\in Z(\mathbf{L})$ and $m^{F'}m^{-1}\in Z(\mathbf{M})$.

The torus $\mathbf{H}^\circ_w$ is large enough for our purpose when one of the Levi subgroups is included in the other one, *e.g.*, when $\mathbf{M}\subset\mathbf{L}$. *From now on we will assume that hypothesis* (beware that in our initial notation this means that ${}^w\mathbf{L}'\subset\mathbf{L}$ for the given element w). To determine $\mathbf{H}^\circ_w$ in that case, we will use the "norm" on a torus.

11.9 NOTATION. *Let* $\mathbf{T}$ *be a torus defined over* $\mathbb{F}_q$ *and let* F *be the corresponding Frobenius endomorphism; given a non-zero integer* $n\in\mathbb{N}$, *we define the morphism*

$$N_{F^n/F} : \mathbf{T} \to \mathbf{T}$$

by $\tau \mapsto \tau.{}^F\tau \dots {}^{F^{n-1}}\tau$.

11.10 LEMMA.

(i) The first projection maps $\mathbf{H}_w^\circ$ *surjectively to* $Z(\mathbf{L})^\circ$.

(ii) If $\mathbf{H}_w^\circ$ *has any fixed point in* $\mathbf{Z}_w''$ *then*

$$\mathbf{H}_w^\circ = \{ (l, l^{-1}) \mid l \in Z(\mathbf{L})^\circ \}.$$

(iii) In general $\mathbf{H}_w^\circ = \{ (N_{F^n/F}(\tau), N_{F'^n/F'}({}^{w'}\tau^{-1})) \mid \tau \in Z(\mathbf{L})^\circ \}$*, where* n *is such that* ${}^{F^n}w = w$ *(for such an* n *we have* $F'^n = F^n$ *and* ${}^{F^n}w' = w'$*); and* $\mathbf{H}_w^\circ \cap (\mathbf{L}^F \times \mathbf{M}^{F'})$ *consists of the pairs such that* $\tau \in (Z(\mathbf{L})^\circ)^{F^n}$.

PROOF: If $l \in Z(\mathbf{L})^\circ$ then ${}^{w'}(l^{-1}{}^F l) \in {}^{w'}Z(\mathbf{L})^\circ = {}^{w'F}Z(\mathbf{L})^\circ \subset {}^{F'}Z(\mathbf{M})^\circ = Z(\mathbf{M})^\circ$. Thus by the Lang-Steinberg theorem there is some $m \in Z(\mathbf{M})^\circ$ such that ${}^{w'}(l^{-1}{}^F l) = m^{F'}m^{-1}$, so the first projection maps $\mathbf{H}_w$ surjectively to $Z(\mathbf{L})^\circ$. As $Z(\mathbf{L})^\circ$ is connected this projection is still surjective when restricted to $\mathbf{H}_w^\circ$, whence (i).

Suppose there exists $(u, u', v, v', n) \in \mathbf{Z}_w''$ fixed by $\mathbf{H}_w^\circ$. Then in particular for any $(l, m) \in \mathbf{H}_w^\circ$ we have $lnm = n$. Since $n \in \mathbf{LM}$ and $l \in Z(\mathbf{L})$, $m \in Z(\mathbf{M})$, we get $lm = 1$, whence the inclusion $\mathbf{H}_w^\circ \subset \{ (l, l^{-1}) \mid l \in Z(\mathbf{L})^\circ \}$. Since the first projection is surjective, this gives (ii).

We now prove (iii). Let $\mathbf{H}_1$ stand for the group that we want $\mathbf{H}_w^\circ$ to be equal to. A straightforward computation shows that $\mathbf{H}_1 \subset \mathbf{H}_w$ (note that ${}^{w'}\tau \in {}^{w'}Z(\mathbf{L})^\circ = Z(\mathbf{M})^\circ$). As $\mathbf{H}_1$ is the image of the connected group $Z(\mathbf{L})^\circ$, it is connected; thus $\mathbf{H}_1 \subset \mathbf{H}_w^\circ$. On the other hand the first projection maps $\mathbf{H}_1$ surjectively to $Z(\mathbf{L})^\circ$ since by the Lang-Steinberg theorem any $t \in Z(\mathbf{L})^\circ$ can be written $s.{}^{F^n}s^{-1}$ with $s \in Z(\mathbf{L})^\circ$. Thus we get $t = N_{F^n/F}(\tau)$ where $\tau = s^F s^{-1}$, and t is the projection of $(N_{F^n/F}(\tau), N_{F'^n/F'}({}^{w'}\tau^{-1}))$. As two elements of $\mathbf{H}_w$ have the same first projection only if their second projections differ by an element of $\mathbf{M}^{F'}$, the group $\mathbf{H}_1$ is of finite index in $\mathbf{H}_w^\circ$, thus is equal to it since it is connected. ■

We note for future reference that, by 10.13 (i), a representation $\pi \otimes \pi' \in \operatorname{Irr}(\mathbf{L}^F \times (\mathbf{M}^{F'})^{\mathrm{opp}})$ can occur in some $H_c^i(\mathbf{Z}_w'')$ only if its restriction to $\mathbf{H}_w^\circ \cap (\mathbf{L}^F \times \mathbf{M}^{F'})$ is trivial.

As stated above the virtual $\mathbf{L}^F$-modules-$\mathbf{M}^{F'}$ given by $H^*(\mathbf{Z}_w'')$ and $H^*(\mathbf{Z}_w''^{\mathbf{H}_w^\circ})$ are isomorphic. By 11.10 (ii), if the element $(u, v, u', v', n) \in \mathbf{Z}_w''$ is fixed by

$\mathbf{H}^\circ_w$ then u, u',v and v' centralize $Z(\mathbf{L})^\circ$. But $C_{\mathbf{G}}(Z(\mathbf{L})^\circ) = \mathbf{L}$ (see 1.21), so this gives $u \in \mathbf{U} \cap \mathbf{L} = \{1\}$, $u' \in {}^{F^{-1}}\mathbf{U} \cap \mathbf{L} = \{1\}$, $v \in \mathbf{V} \cap \mathbf{L}$ and $v' \in {}^{F'^{-1}}\mathbf{V} \cap \mathbf{L}$. So the elements of $(\mathbf{Z}''_w)^{\mathbf{H}^\circ_w}$ are of the form $(1, v, 1, v', n)$ with ${}^F n = nv'vw'$, which proves that $w' \in \mathbf{L}$, since $n \in \mathbf{L}.\mathbf{M} = \mathbf{L}$. Since w is determined up to left multiplication by an element of $\mathbf{L}$, we may assume that $w' = 1$ and thus $F' = F$. We thus get

$$(\mathbf{Z}''_w)^{\mathbf{H}^\circ_w} \simeq \{\, (v, v', n) \in (\mathbf{V} \cap \mathbf{L}) \times ({}^{F^{-1}}\mathbf{V} \cap \mathbf{L}) \times \mathbf{L} \mid {}^F n = nv'v \,\},$$

on which the action of $(l, m) \in \mathbf{L}^F \times \mathbf{M}^F$ is given by

$$(v, v', n) \mapsto ({}^{m^{-1}}v, {}^{m^{-1}}v', lnm).$$

We now look at a term $R^{\mathbf{L}}_{\mathbf{L}\cap{}^w\mathbf{L}'\subset\mathbf{L}\cap{}^w\mathbf{P}'} \circ {}^*R^{{}^w\mathbf{L}'}_{\mathbf{L}\cap{}^w\mathbf{L}'\subset\mathbf{P}\cap{}^w\mathbf{L}'} \circ \mathrm{ad}\, w$ of the right-hand side of the Mackey formula indexed by some $w \in \mathcal{S}(\mathbf{L}, \mathbf{L}')^F$ (we make no particular assumption on $\mathbf{L}$ and $\mathbf{L}'$ until the end of the current paragraph). As for the left-hand side, we first write the $\mathbf{L}^F$-module-$({}^w\mathbf{L}')^F$ to which this functor is associated as $H^*_c(\mathcal{L}^{-1}_{\mathbf{L}}(\mathbf{L} \cap {}^w\mathbf{U}') \times_{(\mathbf{L}\cap{}^w\mathbf{L}')^F} \mathcal{L}^{-1}_{{}^w\mathbf{L}'}(\mathbf{U} \cap {}^w\mathbf{L}'))$. As before, using $\mathbf{V} = {}^w\mathbf{U}'$, $\mathbf{M} = {}^w\mathbf{L}'$ and $\mathbf{Q} = {}^w\mathbf{P}'$ we may "forget w"; the module we study is thus the cohomology of the variety

$$\mathbf{S}_w = \{(l, m) \in \mathbf{L} \times \mathbf{M} \mid l^{-1}.{}^F l \in \mathbf{L} \cap \mathbf{V}, m^{-1}.{}^F m \in \mathbf{U} \cap \mathbf{M}\}/\mathbf{L}^F \cap \mathbf{M}^F$$

on which $(\lambda, \mu) \in \mathbf{L}^F \times \mathbf{M}^F$ acts by $(l, m) \mapsto (\lambda l, \mu^{-1}m)$.

The next lemma shows that the terms corresponding to the same w on both sides of the Mackey formula are equal when ${}^w\mathbf{L}'(= \mathbf{M}) \subset \mathbf{L}$.

11.12 LEMMA. *Under the above hypothesis ($\mathbf{M} \subset \mathbf{L}$) the cohomology spaces $\oplus_i H^i_c(\mathbf{S}_w)$ and $\oplus_i H^i_c({\mathbf{Z}''_w}^{\mathbf{H}^\circ_w})$ are isomorphic as $\mathbf{L}^F$-modules-$\mathbf{M}^F$.*

PROOF: We show first that the map φ from

$$(\mathbf{Z}''_w)^{\mathbf{H}^\circ_w} \simeq \{\, (v, v', n) \in (\mathbf{V} \cap \mathbf{L}) \times ({}^{F^{-1}}\mathbf{V} \cap \mathbf{L}) \times \mathbf{L} \mid {}^F n = nv'v \,\}$$

to $\mathcal{L}^{-1}_{\mathbf{L}}(\mathbf{V} \cap \mathbf{L})$ given by $(v, v', n) \mapsto nv'$ is surjective, with all its fibres isomorphic to the same affine space. Indeed, the image of φ is where we claim it is since $nv' \in \mathbf{L}$ and $\mathcal{L}(nv') = v'^{-1}n^{-1F}n^F v' = v^F v' \in \mathbf{V} \cap \mathbf{L}$; and if $n_1 \in \mathcal{L}^{-1}_{\mathbf{L}}(\mathbf{V} \cap \mathbf{L})$, then

$$\varphi^{-1}(n_1) = \{\, (n_1^{-1F}n_1{}^F v'^{-1}, v', n_1 v'^{-1}) \mid v' \in {}^{F^{-1}}\mathbf{V} \cap \mathbf{L} \,\},$$

which is isomorphic to the affine space ${}^{F^{-1}}\mathbf{V} \cap \mathbf{L}$.

The $\mathbf{L}^F$-action-$\mathbf{M}^F$ on $\mathbf{Z}''_w$ is clearly mapped by φ to the natural $\mathbf{L}^F$-action-$\mathbf{M}^F$ on $\mathcal{L}^{-1}_{\mathbf{L}}(\mathbf{L} \cap \mathbf{V})$ (remember that $\mathbf{M} \subset \mathbf{L}$), and thus for that action the cohomology spaces of $\mathbf{Z}''_w$ and $\mathcal{L}^{-1}_{\mathbf{L}}(\mathbf{L}\cap\mathbf{V})$ are isomorphic as $\mathbf{L}^F$-modules-$\mathbf{M}^F$.

But $\mathbf{M} \subset \mathbf{L}$ also gives $\mathbf{M} \cap \mathbf{U} = \{1\}$ whence

$$\mathbf{S}_w \simeq \{ (l,m) \in \mathbf{L} \times \mathbf{M}^F \mid l^{-1}{}^F l \in \mathbf{L} \cap \mathbf{V} \}/\mathbf{M}^F \simeq \{ l \in \mathbf{L} \mid l^{-1}{}^F l \in \mathbf{L} \cap \mathbf{V} \},$$

which gives the result. ∎

From lemmas 11.7 to 11.12 we get in particular

11.13 THEOREM. *The Mackey formula*

$$^*R^{\mathbf{G}}_{\mathbf{L}\subset\mathbf{P}} \circ R^{\mathbf{G}}_{\mathbf{L}'\subset\mathbf{P}'} = \sum_{w \in \mathbf{L}^F \backslash \mathcal{S}(\mathbf{L},\mathbf{L}')^F/\mathbf{L}'^F} R^{\mathbf{L}}_{\mathbf{L}\cap {}^w\mathbf{L}' \subset \mathbf{L}\cap {}^w\mathbf{P}'} \circ {}^*R^{{}^w\mathbf{L}'}_{\mathbf{L}\cap {}^w\mathbf{L}' \subset \mathbf{P}\cap {}^w\mathbf{L}'} \circ \operatorname{ad} w$$

holds when either $\mathbf{L}$ *or* $\mathbf{L}'$ *is a maximal torus.*

PROOF: If $\mathbf{L}'$ is a torus, then ${}^w\mathbf{L}' \subset \mathbf{L}$ for any $w \in \mathcal{S}(\mathbf{L},\mathbf{L}')$ and all the preceding lemmas hold unconditionally. Since the adjoint of the Mackey formula is the same formula with $\mathbf{L}$ and $\mathbf{L}'$ exchanged, the Mackey formula also holds when $\mathbf{L}$ is a maximal torus, since in that case its adjoint holds by the same argument as above. ∎

As we remarked in chapter 6, when the Mackey formula holds, it implies that $R^{\mathbf{G}}_{\mathbf{L}\subset\mathbf{P}}$ does not depend on $\mathbf{P}$. In what follows we will usually omit $\mathbf{P}$ from the notation; we leave it to the reader to check that we do that in cases where either the choice of a parabolic subgroup is irrelevant or the Mackey formula holds.

11.14 DEFINITION. *When* $\mathbf{L} = \mathbf{T}$ *is a rational maximal torus,* $R^{\mathbf{G}}_{\mathbf{T}}(\theta)$ *for* $\theta \in \operatorname{Irr}(\mathbf{T}^F)$ *is called a* **Deligne-Lusztig character**.

These characters were introduced in [DL1]. The Mackey formula gives the "scalar product formula for Deligne-Lusztig characters".

11.15 COROLLARY. *Let* $\mathbf{T}$ *and* $\mathbf{T}'$ *be two rational maximal tori. Then:*

(i) *For* $\theta \in \operatorname{Irr}(\mathbf{T}^F)$ *and* $\theta' \in \operatorname{Irr}(\mathbf{T}'^F)$ *we have*

$$\langle R^{\mathbf{G}}_{\mathbf{T}}(\theta), R^{\mathbf{G}}_{\mathbf{T}'}(\theta') \rangle_{\mathbf{G}^F} = |\mathbf{T}^F|^{-1} \#\{n \in \mathbf{G}^F \mid {}^n\mathbf{T} = \mathbf{T}' \text{ and } {}^n\theta = \theta'\}.$$

(ii) *The functor* $R^{\mathbf{G}}_{\mathbf{T}\subset\mathbf{B}}$ *does not depend on the Borel subgroup* $\mathbf{B}$ *used in its construction.*

PROOF: (i) is just another way of writing the Mackey formula, and (ii) is clear from the above remarks. ∎

We remark that, as $R^{\mathbf{G}}_{\mathbf{T}}(\theta) = R^{\mathbf{G}}_{\mathbf{T}'}(\theta')$ when ${}^g(\mathbf{T},\theta) = (\mathbf{T}',\theta')$ for some $g \in \mathbf{G}^F$, (i) above shows that the $R^{\mathbf{G}}_{\mathbf{T}}(\theta)$ give a set of orthogonal class functions indexed by the $\mathbf{G}^F$-conjugacy classes of pairs $(\mathbf{T},\theta)$.

In 13.3, in connection with Lusztig series, we will give a statement somewhat stronger than 11.15 which deals directly with the cohomology groups of $\mathcal{L}^{-1}(\mathbf{U})$, where $\mathbf{B} = \mathbf{TU}$. The next corollary gives 11.15 (i) in terms of Weyl groups when the characters θ and θ' are the identity.

11.16 COROLLARY. *Given a rational maximal torus* $\mathbf{T}$ *with Weyl group* W, *let* $\mathbf{T}_w$ *(resp.* $\mathbf{T}_{w'}$*) be a rational maximal torus of type* w *(resp.* w'*) with respect to* $\mathbf{T}$ *(see 3.24); then*

$$\langle R^{\mathbf{G}}_{\mathbf{T}_w}(\mathrm{Id}_{\mathbf{T}_w}), R^{\mathbf{G}}_{\mathbf{T}_{w'}}(\mathrm{Id}_{{}^bT_{w'}}) \rangle_{\mathbf{G}^F} = \begin{cases} |W^{wF}| & \text{if } w \text{ and } w' \text{ are } F\text{-conjugate in } W, \\ 0 & \text{otherwise.} \end{cases}$$

PROOF: The scalar product is 0 unless $\mathbf{T}_w$ and $\mathbf{T}_{w'}$ are $\mathbf{G}^F$-conjugate, *i.e.*, unless w and w' are F-conjugate (see 3.23). In this last case we may assume $w = w'$ and $\mathbf{T}_w = \mathbf{T}_{w'}$. Then 11.15 gives $\langle R^{\mathbf{G}}_{\mathbf{T}_w}(\mathrm{Id}_{\mathbf{T}_w}), R^{\mathbf{G}}_{\mathbf{T}_w}(\mathrm{Id}_{\mathbf{T}_w}) \rangle_{\mathbf{G}^F} = |\mathbf{T}_w^F|^{-1}|N_{\mathbf{G}^F}(\mathbf{T}_w)| = |W(\mathbf{T}_w)^F|$. As the action of F on the Weyl group of $\mathbf{T}_w$ can be identified with the action of wF on W, we get the result. ■

References

The construction of $R^{\mathbf{G}}_{\mathbf{T}}$ is one of the fundamental ideas of [DL1]. The construction of $R^{\mathbf{G}}_{\mathbf{L}}$ was first published in [L1] as a natural extension of that construction. The Mackey formula when one of the Levi subgroups is a torus is given in [DL2 II, theorem 7], but the proof in that paper has an error. The proof we give here corrects this error, using an argument indicated to us by Lusztig. The case of two tori (11.15) was already in [DL1].

12. THE CHARACTER FORMULA AND OTHER RESULTS ON DELIGNE-LUSZTIG INDUCTION.

In this chapter we give the character formulae for Deligne-Lusztig induction and restriction. We then use it to generalize the results of chapter 7, as well as some results of chapters 8 and 9, and we express the identity, Steinberg and regular characters, and the characteristic function of a semi-simple conjugacy class, as linear combinations of Deligne-Lusztig characters.

12.1 DEFINITION. *Given* $\mathbf{L}$, *a rational Levi subgroup of* $\mathbf{G}$ *and a Levi decomposition* $\mathbf{P} = \mathbf{L}\mathbf{U}$ *of a (possibly non-rational) parabolic subgroup containing* $\mathbf{L}$, *the (two-variable)* **Green function** $Q^{\mathbf{G}}_{\mathbf{L}\subset\mathbf{P}} : \mathbf{G}^F_u \times \mathbf{L}^F_u \to \mathbb{Z}$ *is defined by*

$$(u, v) \mapsto |\mathbf{L}^F|^{-1}\,\mathrm{Trace}((u, v) \mid H^*_c(\mathcal{L}^{-1}(\mathbf{U}))).$$

We recall that we use the notation $\mathbf{G}_u$ to denote the set of unipotent elements of an algebraic group $\mathbf{G}$. The values of the Green function are in $\mathbb{Z}$ by 10.6. As explained for $R^{\mathbf{G}}_{\mathbf{L}}$, we usually omit $\mathbf{P}$ from the notation.

12.2 PROPOSITION (CHARACTER FORMULA FOR $R^{\mathbf{G}}_{\mathbf{L}}$ AND ${}^*R^{\mathbf{G}}_{\mathbf{L}}$). *Let* $\mathbf{L}$ *be a rational Levi subgroup of* $\mathbf{G}$ *and let* $\psi \in \mathrm{Irr}(\mathbf{G}^F)$ *and* $\chi \in \mathrm{Irr}(\mathbf{L}^F)$, *then*

$$(R^{\mathbf{G}}_{\mathbf{L}}\chi)(g) = \\ |\mathbf{L}^F|^{-1}|C^\circ_{\mathbf{G}}(s)^F|^{-1} \sum_{\{h\in\mathbf{G}^F \mid s\in {}^h\mathbf{L}\}} |C^\circ_{{}^h\mathbf{L}}(s)^F| \sum_{v\in C^\circ_{{}^h\mathbf{L}}(s)^F_u} Q^{C^\circ_{\mathbf{G}}(s)}_{C^\circ_{{}^h\mathbf{L}}(s)}(u, v^{-1})\,{}^h\chi(sv) \quad \text{(i)}$$

where $g = su$ *is the Jordan decomposition of* $g \in \mathbf{G}^F$, *and*

$$({}^*R^{\mathbf{G}}_{\mathbf{L}}\psi)(l) = |C^\circ_{\mathbf{L}}(t)^F||C^\circ_{\mathbf{G}}(t)^F|^{-1} \sum_{u\in C^\circ_{\mathbf{G}}(t)^F_u} Q^{C^\circ_{\mathbf{G}}(t)}_{C^\circ_{\mathbf{L}}(t)}(u, v^{-1})\psi(tu) \quad \text{(ii)}$$

where $l = tv$ *is the Jordan decomposition of* $l \in \mathbf{L}^F$.

PROOF: The main step is the following lemma.

12.3 LEMMA. *With the above notation, we have*

$$\mathrm{Trace}((g, l) \mid H^*_c(\mathcal{L}^{-1}(\mathbf{U}))) \\ = |C^\circ_{\mathbf{L}}(t)^F||C^\circ_{\mathbf{G}}(t)^F|^{-1} \sum_{\{h\in\mathbf{G}^F \mid {}^ht=s^{-1}\}} Q^{C^\circ_{\mathbf{G}}(t)}_{C^\circ_{\mathbf{L}}(t)}({}^{h^{-1}}u, v) \quad (*) \\ = |C^\circ_{\mathbf{G}}(s)^F|^{-1} \sum_{\{h\in\mathbf{G}^F \mid {}^{h^{-1}}s=t^{-1}\}} |C^\circ_{{}^h\mathbf{L}}(s)^F| Q^{C^\circ_{\mathbf{G}}(s)}_{C^\circ_{{}^h\mathbf{L}}(s)}(u, {}^hv). \quad (**)$$

PROOF: These two equalities are clearly equivalent; we shall prove the first one. From 10.14 we get

$$\mathrm{Trace}((g,l)\mid H^*_c(\mathcal{L}^{-1}(\mathbf{U}))) = \mathrm{Trace}((u,v)\mid H^*_c(\mathcal{L}^{-1}(\mathbf{U})^{(s,t)})).$$

We first show that the morphism

$$\varphi : \{\, h \in \mathbf{G}^F \mid {}^h t = s^{-1} \,\} \times \{\, z \in C^\circ_{\mathbf{G}}(t) \mid z^{-1}\,{}^F z \in \mathbf{U} \,\} \to \mathcal{L}^{-1}(\mathbf{U})^{(s,t)}$$

given by $(h,z) \mapsto hz$ is surjective (it is easy to check that the image of φ is in $\mathcal{L}^{-1}(\mathbf{U})^{(s,t)}$ since $\mathcal{L}^{-1}(\mathbf{U})^{(s,t)} = \{x \in \mathbf{G} \mid x^{-1}\,{}^F x \in \mathbf{U}$ and $sxt = x\}$). If $x \in \mathcal{L}^{-1}(\mathbf{U})^{(s,t)}$ then $sxt = x$ implies $s^F xt = {}^F x = x(x^{-1}\,{}^F x)$ whence $s^F xt = sxt(x^{-1}\,{}^F x)$, which can be written $(x^{-1}\,{}^F x)t = t(x^{-1}\,{}^F x)$, *i.e.*, $x^{-1}\,{}^F x \in C_{\mathbf{G}}(t)$. As $x^{-1}\,{}^F x$ is unipotent, we even get $x^{-1}\,{}^F x \in C^\circ_{\mathbf{G}}(t)$ (see 2.5). Applying the Lang-Steinberg theorem in the group $C^\circ_{\mathbf{G}}(t)$ we may write $x^{-1}\,{}^F x = z^{-1}\,{}^F z$, where $z \in C^\circ_{\mathbf{G}}(t)$. If we put then $h = xz^{-1}$ we have $h \in \mathbf{G}^F$, ${}^h t = s^{-1}$ and $\varphi(h,z) = x$, whence the surjectivity of φ.

The map φ is not injective, but $\varphi(h,z) = \varphi(h',z')$ if and only if $h^{-1}h' = zz'^{-1} \in C^\circ_{\mathbf{G}}(t)^F$, thus φ induces an isomorphism

$$\{\, h \in \mathbf{G}^F \mid {}^h t = s^{-1} \,\} \times_{C^\circ_{\mathbf{G}}(t)^F} \{\, z \in C^\circ_{\mathbf{G}}(t) \mid z^{-1}\,{}^F z \in \mathbf{U} \,\} \xrightarrow{\sim} \mathcal{L}^{-1}(\mathbf{U})^{(s,t)},$$

which may be written

$$\begin{aligned}\mathcal{L}^{-1}(\mathbf{U})^{(s,t)} &\simeq \coprod_{\{\, h\in \mathbf{G}^F/C^\circ_{\mathbf{G}}(t)^F \mid {}^h t = s^{-1}\,\}} \{\, z \in C^\circ_{\mathbf{G}}(t) \mid z^{-1}\,{}^F z \in \mathbf{U} \,\}_h \\ &= \coprod_{\{\, h\in \mathbf{G}^F/C^\circ_{\mathbf{G}}(t)^F \mid {}^h t = s^{-1}\,\}} \mathcal{L}^{-1}_{C^\circ_{\mathbf{G}}(t)}(\mathbf{U} \cap C^\circ_{\mathbf{G}}(t))_h,\end{aligned}$$

where $(u,v) \in C^\circ_{\mathbf{G}}(s)^F_u \times C^\circ_{\mathbf{L}}(t)^F_u$ acts on the piece indexed by h by $z \mapsto {}^{h^{-1}}uzv$. We thus get

$$\mathrm{Trace}((g,l)\mid H^*_c(\mathcal{L}^{-1}(\mathbf{U}))) = |C^\circ_{\mathbf{G}}(t)^F|^{-1} \sum_{\{\, h\in \mathbf{G}^F \mid {}^h t = s^{-1}\,\}} \mathrm{Trace}(({}^{h^{-1}}u, v)\mid H^*_c(\mathcal{L}^{-1}_{C^\circ_{\mathbf{G}}(t)}(\mathbf{U} \cap C^\circ_{\mathbf{G}}(t))),$$

whence the lemma. ∎

We now prove proposition 12.2. From proposition 11.2 we have

$$(R^{\mathbf{G}}_{\mathbf{L}}\chi)(g) = |\mathbf{L}^F|^{-1} \sum_{l\in \mathbf{L}^F} \mathrm{Trace}((g,l)\mid H^*_c(\mathcal{L}^{-1}(\mathbf{U})))\chi(l^{-1})$$

and

$$({}^*R_{\mathbf{L}}^{\mathbf{G}}\psi)(l) = |\mathbf{G}^F|^{-1} \sum_{g\in\mathbf{G}^F} \mathrm{Trace}((g,l) \mid H_c^*(\mathcal{L}^{-1}(\mathbf{U})))\psi(g^{-1}).$$

Applying lemma 12.3 in both sums, and interchanging the sums, we get from the first formula and (**)

$$(R_{\mathbf{L}}^{\mathbf{G}}\chi)(g) =$$
$$|\mathbf{L}^F|^{-1}|C_{\mathbf{G}}^\circ(s)^F|^{-1} \sum_{\{h\in\mathbf{G}^F|^{h^{-1}}s\in\mathbf{L}^F\}} |C_{{}^h\mathbf{L}}^\circ(s)^F| \sum_{v\in C_{\mathbf{L}}^\circ({}^{h^{-1}}s)_u^F} Q_{C_{{}^h\mathbf{L}}^\circ(s)}^{C_{\mathbf{G}}^\circ(s)}(u, {}^hv)\chi({}^{h^{-1}}sv^{-1})$$

and from the second formula and (*)

$$({}^*R_{\mathbf{L}}^{\mathbf{G}}\psi)(l) = |\mathbf{G}^F|^{-1}|C_{\mathbf{G}}^\circ(t)^F|^{-1}|C_{\mathbf{L}}^\circ(t)^F| \sum_{h\in\mathbf{G}^F} \sum_{u\in C_{\mathbf{G}}^\circ({}^ht)_u^F} Q_{C_{\mathbf{L}}^\circ(t)}^{C_{\mathbf{G}}^\circ(t)}({}^{h^{-1}}u, v)\psi({}^htu^{-1}).$$

We then get the result by changing the variable on which we sum to ${}^hv^{-1}$ in the first formula and to ${}^{h^{-1}}u^{-1}$ in the second. ■

12.4 COROLLARY. *Under the same assumptions as proposition 12.2, we have*

$$(R_{\mathbf{L}}^{\mathbf{G}}\chi)(g) = |C_{\mathbf{G}}^\circ(s)^F|^{-1} \sum_{\{h\in\mathbf{G}^F|s\in{}^h\mathbf{L}\}} |C_{{}^h\mathbf{L}}^\circ(s)^F||\mathbf{L}^F|^{-1} R_{C_{{}^h\mathbf{L}}^\circ(s)}^{C_{\mathbf{G}}^\circ(s)}({}^h\chi)(g).$$

PROOF: In the case where $s \in Z(\mathbf{G})$, lemma 12.3 (*) reduces to

$$|\mathbf{L}^F|^{-1}\,\mathrm{Trace}((g,l) \mid H_c^*(\mathcal{L}^{-1}(\mathbf{U})))$$
$$= \begin{cases} |\mathbf{G}^F|^{-1}\sum_{h\in\mathbf{G}^F} Q_{\mathbf{L}}^{\mathbf{G}}({}^{h^{-1}}u, v) = Q_{\mathbf{L}}^{\mathbf{G}}(u,v) & \text{if } s = t^{-1}, \\ 0 & \text{otherwise,} \end{cases}$$

where the last equality in the first line comes from the fact that $Q_{\mathbf{L}}^{\mathbf{G}}$ is the restriction to unipotent elements of a central function on $\mathbf{G}^F \times \mathbf{L}^F$. So in that case 11.2 gives

$$(R_{\mathbf{L}}^{\mathbf{G}}\chi)(g) = \sum_{v\in\mathbf{L}_u^F} Q_{\mathbf{L}}^{\mathbf{G}}(u, v^{-1})\chi(sv).$$

Applying this formula to $R_{C_{{}^h\mathbf{L}}^\circ(s)}^{C_{\mathbf{G}}^\circ(s)}({}^h\chi)(g)$ in the right-hand side of the equality we want to prove, we see that it is equivalent to 12.2 (i). ■

We now use the character formula to extend 7.6 to the Deligne-Lusztig induction.

12.5 PROPOSITION. *Let* $\mathbf{L}$ *be a rational Levi subgroup of* $\mathbf{G}$, *let* $\psi \in \operatorname{Irr}(\mathbf{G}^F)$ *and let* $l = su$ *be the Jordan decomposition of some element* $l \in \mathbf{L}^F$. *Then*

$$((\operatorname{Res}^{\mathbf{L}^F}_{C^\circ_{\mathbf{L}}(s)^F} \circ {}^*R^{\mathbf{G}}_{\mathbf{L}})\psi)(l) = (({}^*R^{C^\circ_{\mathbf{G}}(s)}_{C^\circ_{\mathbf{L}}(s)} \circ \operatorname{Res}^{\mathbf{G}^F}_{C^\circ_{\mathbf{G}}(s)^F})\psi)(l).$$

PROOF: This results from the remark that, in the character formula 12.2 for ${}^*R^{\mathbf{G}}_{\mathbf{L}}\psi$, the right-hand side does not change if we replace $\mathbf{G}$ by $C^\circ_{\mathbf{G}}(s)$ and $\mathbf{L}$ by $C^\circ_{\mathbf{L}}(s)$. ∎

The next proposition and its corollary extend 7.4 and 7.5.

12.6 PROPOSITION. *Let* $f \in \mathcal{C}(\mathbf{G}^F)_{p'}$; *then, for any rational Levi subgroup* $\mathbf{L}$ *of* $\mathbf{G}$ *and any* $\pi \in \mathcal{C}(\mathbf{L}^F)$ *(resp.* $\psi \in \mathcal{C}(\mathbf{G}^F)$*), we have*

$$R^{\mathbf{G}}_{\mathbf{L}}(\pi . \operatorname{Res}^{\mathbf{G}^F}_{\mathbf{L}^F} f) = (R^{\mathbf{G}}_{\mathbf{L}}\pi).f, \qquad (i)$$

$$({}^*R^{\mathbf{G}}_{\mathbf{L}}\psi). \operatorname{Res}^{\mathbf{G}^F}_{\mathbf{L}^F} f = {}^*R^{\mathbf{G}}_{\mathbf{L}}(\psi.f). \qquad (ii)$$

PROOF: Proposition 12.2 gives

$$R^{\mathbf{G}}_{\mathbf{L}}(\pi . \operatorname{Res}^{\mathbf{G}^F}_{\mathbf{L}^F} f)(g) =$$
$$|\mathbf{L}^F|^{-1}|C^\circ_{\mathbf{G}}(s)^F|^{-1} \sum_{\{h\in\mathbf{G}^F \mid s\in {}^h\mathbf{L}\}} |C^\circ_{{}^h\mathbf{L}}(s)^F| \sum_{v\in C^\circ_{{}^h\mathbf{L}}(s)^F_u} Q^{C^\circ_{\mathbf{G}}(s)}_{C^\circ_{{}^h\mathbf{L}}(s)}(u, v^{-1})\, {}^h\pi(sv)\, {}^hf(sv),$$

which gives (i) using ${}^hf(sv) = f(sv) = f(s) = f(g)$; equality (ii) is proved similarly (it can also be obtained from (i) by adjunction). ∎

12.7 COROLLARY. *Under the same assumptions, we have* ${}^*R^{\mathbf{G}}_{\mathbf{L}} f = \operatorname{Res}^{\mathbf{G}^F}_{\mathbf{L}^F} f$.

PROOF: This results from the special case of 12.6 where $\psi = \operatorname{Id}_{\mathbf{G}}$, and the remark that ${}^*R^{\mathbf{G}}_{\mathbf{L}}(\operatorname{Id}_{\mathbf{G}}) = \operatorname{Id}_{\mathbf{L}}$. Let us prove this last fact: by definition ${}^*R^{\mathbf{G}}_{\mathbf{L}}(\operatorname{Id}_{\mathbf{G}})$ is the character afforded by the $\mathbf{L}^F$-module $H^*_c(\mathcal{L}^{-1}(\mathbf{U}))^{\mathbf{G}^F}$; by 10.10 this module is isomorphic to $H^*_c(\mathcal{L}^{-1}(\mathbf{U})/\mathbf{G}^F)$, and the Lang map induces an isomorphism from $\mathcal{L}^{-1}(\mathbf{U})/\mathbf{G}^F$ to $\mathbf{U}$, whence the result by 10.11. ∎

As discussed in the proof of 8.11 the validity of the Mackey formula when one of the Levi subgroups is a torus (see 11.13) allows us to state the following analogue of 8.11 (using the fact that the duality in a torus is the identity).

12.8 THEOREM. *For any rational maximal torus* $\mathbf{T}$ *of* $\mathbf{G}$ *we have*

$$\varepsilon_{\mathbf{G}} D_{\mathbf{G}} \circ R^{\mathbf{G}}_{\mathbf{T}} = \varepsilon_{\mathbf{T}} R^{\mathbf{G}}_{\mathbf{T}}.$$

As in the beginning of chapter 9 we immediately deduce (using the value of ${}^*R^{\mathbf{G}}_{\mathbf{T}}(\operatorname{Id}_{\mathbf{G}})$ given in the proof of 12.7 instead of 7.4) that

$${}^*R^{\mathbf{G}}_{\mathbf{T}} \operatorname{St}_{\mathbf{G}} = \varepsilon_{\mathbf{G}}\varepsilon_{\mathbf{T}} \operatorname{St}_{\mathbf{T}} = \varepsilon_{\mathbf{G}}\varepsilon_{\mathbf{T}} \operatorname{Id}_{\mathbf{T}}.$$

We can now give the dimension of the (virtual) characters $R^{\mathbf{G}}_{\mathbf{T}}(\theta)$.

12.9 PROPOSITION. *For any rational maximal torus* $\mathbf{T}$ *and any* $\theta \in \operatorname{Irr}(\mathbf{T}^F)$, *we have* $\dim R_{\mathbf{T}}^{\mathbf{G}}(\theta) = \varepsilon_{\mathbf{G}}\varepsilon_{\mathbf{T}}|\mathbf{G}^F|_{p'}|\mathbf{T}^F|^{-1}$.

PROOF: By 12.2 the dimension we want to compute does not depend on θ. On the other hand, taking the scalar product of the equalities given above with θ, we get $\langle R_{\mathbf{T}}^{\mathbf{G}}(\theta), \mathrm{St}_{\mathbf{G}}\rangle_{\mathbf{G}^F} = \varepsilon_{\mathbf{T}}\varepsilon_{\mathbf{G}}\delta_{1,\theta}$, whence $\langle \sum_\theta R_{\mathbf{T}}^{\mathbf{G}}(\theta), \mathrm{St}_{\mathbf{G}}\rangle_{\mathbf{G}^F} = \varepsilon_{\mathbf{T}}\varepsilon_{\mathbf{G}}$. But $\sum_\theta R_{\mathbf{T}}^{\mathbf{G}}(\theta)$ is the character afforded by the module $H_c^*(\mathcal{L}^{-1}(\mathbf{U}))$, where $\mathbf{U}$ is the unipotent radical of some Borel subgroup containing $\mathbf{T}$. By 10.14 this character vanishes on all non-trivial semi-simple elements, as these elements have no fixed points on $\mathcal{L}^{-1}(\mathbf{U})$. Since $\mathrm{St}_{\mathbf{G}}$ vanishes outside semi-simple elements, the scalar product above reduces to

$$|\mathbf{G}^F|^{-1}|\mathbf{T}^F|\,\mathrm{St}_{\mathbf{G}}(1)\dim(R_{\mathbf{T}}^{\mathbf{G}}(\theta)).$$

This gives the result after replacing $\mathrm{St}_{\mathbf{G}}(1)$ by its value. ■

12.10 REMARK. *If* $\mathbf{T}_w$ *is a maximal rational torus of type* $w \in W(\mathbf{T})$ *with respect to some quasi-split torus* $\mathbf{T}$, *we have* $\varepsilon_{\mathbf{T}_w}\varepsilon_{\mathbf{G}} = (-1)^{l(w)}$, *where* $l(w)$ *is the length of* w *in* $W(\mathbf{T})$.

PROOF: Note first that the statement makes sense since $(-1)^{l(w)}$ is defined independently not only of the choice of a set of generating reflections in W (since such a set corresponds to a Borel subgroup containing $\mathbf{T}$, so two such sets are conjugate by an element of W, and $(-1)^{l(vwv^{-1})} = (-1)^{l(w)}$) but also of the chosen type of $\mathbf{T}_w$ (since $(-1)^{l(vw.{}^Fv^{-1})} = (-1)^{l(w)}$ because $l({}^Fv) = l(v)$). Let $E = X(\mathbf{T})\otimes\mathbb{R}$; with the notation of 8.2 we have $\varepsilon_{\mathbf{G}} = (-1)^{\dim(E^\tau)}$ and, as $\mathbf{T}_w$ with the action of F can be identified to $\mathbf{T}$ with the action of wF (see 3.24), we have $\varepsilon_{\mathbf{T}_w} = (-1)^{\dim(E^{w\tau})}$. Since τ is an automorphism of finite order of the lattice $X(\mathbf{T})$, we have $(-1)^{\dim(E)-\dim(E^\tau)} = \det(\tau)$, and similarly $(-1)^{\dim(E)-\dim(E^{w\tau})} = \det(w\tau)$, which gives the result as $\det(w) = (-1)^{l(w)}$, since the determinant of a reflection is -1. ■

To get further properties of the Lusztig functor, in particular the dimension of $R_{\mathbf{L}}^{\mathbf{G}}(\chi)$ and the analogue of 9.6, we first need to prove that the identity and the regular representation are both linear combinations of Deligne-Lusztig characters. We will use the following terminology.

12.11 DEFINITION. *We call* **uniform functions** *the class functions on* $\mathbf{G}^F$ *that are linear combinations of Deligne-Lusztig characters.*

In the formulae up to the end of this chapter, we will let $\mathcal{T}$ denote the set of all rational maximal tori of $\mathbf{G}$, and $[\mathcal{T}/\mathbf{G}^F]$ denote a set of representatives of $\mathbf{G}^F$-conjugacy classes of maximal rational tori. We fix $\mathbf{T}_1 \in \mathcal{T}$ and we put $W = W(\mathbf{T}_1)$. We recall that $\mathcal{T}/\mathbf{G}^F$ is in one-to-one correspondence with the F-classes of W (see 3.23). Finally for each $w \in W$ we choose some rational maximal torus $\mathbf{T}_w$ of type w with respect to $\mathbf{T}_1$.

12.12 PROPOSITION. *The orthogonal projection of class functions onto the subspace of uniform functions is given by the operator:*

$$p = |W|^{-1} \sum_{w\in W} R^{\mathbf{G}}_{\mathbf{T}_w} \circ {}^*R^{\mathbf{G}}_{\mathbf{T}_w} = \sum_{\mathbf{T}\in[\mathcal{T}/\mathbf{G}^F]} |W(\mathbf{T})^F|^{-1} R^{\mathbf{G}}_{\mathbf{T}} \circ {}^*R^{\mathbf{G}}_{\mathbf{T}}$$
$$= |\mathbf{G}^F|^{-1} \sum_{\mathbf{T}\in\mathcal{T}} |\mathbf{T}^F| R^{\mathbf{G}}_{\mathbf{T}} \circ {}^*R^{\mathbf{G}}_{\mathbf{T}}$$

PROOF: The equality of the three expressions for p results from a straightforward computation. Let us check that the middle one is a projector on uniform functions. Since $p(\chi)$ is clearly uniform for any $\chi \in \operatorname{Irr}(\mathbf{G}^F)$, it is enough to check that for any $\mathbf{T} \in [\mathcal{T}/\mathbf{G}^F]$ and any $\theta \in \operatorname{Irr}(\mathbf{T}^F)$, we have $\langle \chi, R^{\mathbf{G}}_{\mathbf{T}}(\theta) \rangle_{\mathbf{G}^F} = \langle p(\chi), R^{\mathbf{G}}_{\mathbf{T}}(\theta) \rangle_{\mathbf{G}^F}$. We have

$$\langle p(\chi), R^{\mathbf{G}}_{\mathbf{T}}(\theta) \rangle_{\mathbf{G}^F} = \langle \sum_{\mathbf{T}'\in[\mathcal{T}/\mathbf{G}^F]} |W(\mathbf{T}')^F|^{-1} R^{\mathbf{G}}_{\mathbf{T}'} {}^*R^{\mathbf{G}}_{\mathbf{T}'} \chi, R^{\mathbf{G}}_{\mathbf{T}}(\theta) \rangle_{\mathbf{G}^F}$$
$$= \sum_{\mathbf{T}'\in[\mathcal{T}/\mathbf{G}^F]} \langle |W(\mathbf{T}')^F|^{-1} {}^*R^{\mathbf{G}}_{\mathbf{T}'} \chi, {}^*R^{\mathbf{G}}_{\mathbf{T}'} R^{\mathbf{G}}_{\mathbf{T}}(\theta) \rangle_{\mathbf{G}^F}$$

but, by 11.15 we have:

$${}^*R^{\mathbf{G}}_{\mathbf{T}'} R^{\mathbf{G}}_{\mathbf{T}}(\theta) = \begin{cases} \sum_{w\in W(\mathbf{T})^F} {}^w\theta & \text{if } \mathbf{T} = \mathbf{T}' \\ 0 & \text{if } \mathbf{T} \text{ and } \mathbf{T}' \text{ are not } \mathbf{G}^F\text{-conjugate} \end{cases}$$

so

$$\langle p(\chi), R^{\mathbf{G}}_{\mathbf{T}}(\theta) \rangle_{\mathbf{G}^F} = \langle {}^*R^{\mathbf{G}}_{\mathbf{T}}(\chi), |W(\mathbf{T})^F|^{-1} \sum_{w\in W(\mathbf{T})^F} {}^w\theta \rangle_{\mathbf{G}^F} = \langle \chi, R^{\mathbf{G}}_{\mathbf{T}}(\theta) \rangle_{\mathbf{G}^F}$$

the rightmost equality since for any θ we have $R^{\mathbf{G}}_{\mathbf{T}}({}^w\theta) = R^{\mathbf{G}}_{\mathbf{T}}(\theta)$. ■

12.13 PROPOSITION. $\operatorname{Id}_{\mathbf{G}}$ *is a uniform function; we have*

$$\operatorname{Id}_{\mathbf{G}} = |W|^{-1} \sum_{w\in W} R^{\mathbf{G}}_{\mathbf{T}_w}(\operatorname{Id}_{\mathbf{T}_w}) = \sum_{\mathbf{T}\in[\mathcal{T}/\mathbf{G}^F]} |W(\mathbf{T})^F|^{-1} R^{\mathbf{G}}_{\mathbf{T}}(\operatorname{Id}_{\mathbf{T}})$$
$$= |\mathbf{G}^F|^{-1} \sum_{\mathbf{T}\in\mathcal{T}} |\mathbf{T}^F| R^{\mathbf{G}}_{\mathbf{T}}(\operatorname{Id}_{\mathbf{T}})$$

PROOF: Since by 12.7 we have ${}^*R^{\mathbf{G}}_{\mathbf{T}}(\operatorname{Id}_{\mathbf{G}}) = \operatorname{Id}_{\mathbf{T}}$, the three expressions above all represent $p(\operatorname{Id}_{\mathbf{G}})$. It is enough to check that $\operatorname{Id}_{\mathbf{G}}$ has same scalar product with one of these expressions as with itself. But indeed we have

$$\langle \operatorname{Id}_{\mathbf{G}}, |W|^{-1} \sum_{w\in W} R^{\mathbf{G}}_{\mathbf{T}_w}(\operatorname{Id}_{\mathbf{T}_w}) \rangle_{\mathbf{G}^F} = |W|^{-1} \sum_{w\in W} \langle {}^*R^{\mathbf{G}}_{\mathbf{T}_w}(\operatorname{Id}_{\mathbf{G}}), \operatorname{Id}_{\mathbf{T}_w} \rangle_{\mathbf{T}_w^F} = 1$$

using again that ${}^*R^{\mathbf{G}}_{\mathbf{T}_w}(\operatorname{Id}_{\mathbf{G}}) = \operatorname{Id}_{\mathbf{T}_w}$. ■

12.14 COROLLARY. *The character* $\mathrm{reg}_{\mathbf{G}}$ *of the regular representation of* $\mathbf{G}^F$ *is a uniform function; we have*

$$\mathrm{reg}_{\mathbf{G}} = |W|^{-1} \sum_{w\in W} \dim(R^{\mathbf{G}}_{\mathbf{T}_w}(\mathrm{Id}_{\mathbf{T}_w}))R^{\mathbf{G}}_{\mathbf{T}_w}(\mathrm{reg}_{\mathbf{T}_w}) = |\mathbf{G}^F|_p^{-1} \sum_{\mathbf{T}\in\mathcal{T}} \varepsilon_{\mathbf{G}}\varepsilon_{\mathbf{T}} R^{\mathbf{G}}_{\mathbf{T}}(\mathrm{reg}_{\mathbf{T}})$$
$$= |\mathbf{G}^F|_p^{-1} \sum_{\substack{\mathbf{T}\in\mathcal{T}\\ \theta\in\mathrm{Irr}(\mathbf{T}^F)}} \varepsilon_{\mathbf{G}}\varepsilon_{\mathbf{T}} R^{\mathbf{G}}_{\mathbf{T}}(\theta).$$

PROOF: Again, the equality of the three expressions is straightforward. Let us get the first one. By 7.4 and 7.5 we have $D_{\mathbf{G}}(\chi f) = D_{\mathbf{G}}(\chi)f$ if $f \in \mathcal{C}(\mathbf{G}^F)_{p'}$; so 9.4 gives $\mathrm{reg}_{\mathbf{G}} = D_{\mathbf{G}}\gamma_p = D_{\mathbf{G}}(\mathrm{Id}_{\mathbf{G}})\gamma_p = \mathrm{St}_{\mathbf{G}}\,\gamma_p$. From 12.13 and 12.8 we get

$$\mathrm{St}_{\mathbf{G}} = D_{\mathbf{G}}(\mathrm{Id}_{\mathbf{G}}) = |W|^{-1} \sum_{w\in W} \varepsilon_{\mathbf{G}}\varepsilon_{\mathbf{T}_w} R^{\mathbf{G}}_{\mathbf{T}_w}(\mathrm{Id}_{\mathbf{T}_w})$$

So it is enough to see that $\varepsilon_{\mathbf{G}}\varepsilon_{\mathbf{T}_w} R^{\mathbf{G}}_{\mathbf{T}_w}(\mathrm{Id}_{\mathbf{T}_w})\gamma_p = \dim R^{\mathbf{G}}_{\mathbf{T}_w}(\mathrm{Id}_{\mathbf{T}})R^{\mathbf{G}}_{\mathbf{T}_w}(\mathrm{reg}_{\mathbf{T}_w})$. This comes from the equality $R^{\mathbf{G}}_{\mathbf{T}_w}(\mathrm{Id}_{\mathbf{T}_w})\gamma_p = R^{\mathbf{G}}_{\mathbf{T}_w}(\mathrm{Res}^{\mathbf{G}^F}_{\mathbf{T}_w^F}(\gamma_p))$ given by 12.6, from the fact that $\mathrm{Res}^{\mathbf{G}^F}_{\mathbf{T}_w^F}(\gamma_p)$ has value $|\mathbf{G}^F|_{p'}$ at 1 and 0 elsewhere, so is equal to $|\mathbf{G}^F|_{p'}|\mathbf{T}_w^F|^{-1}\mathrm{reg}_{\mathbf{T}_w}$, and from 12.9. ■

12.15 COROLLARY. *The number of rational maximal tori of* $\mathbf{G}$ *is equal to* $|\mathbf{G}^F|_p^2$.

PROOF: For any rational maximal torus $\mathbf{T}$, we have $\mathrm{Ind}^{\mathbf{G}^F}_{\mathbf{T}^F}(\mathrm{reg}_{\mathbf{T}}) = \mathrm{reg}_{\mathbf{G}} = |\mathcal{T}|^{-1}\sum_{\mathbf{T}\in\mathcal{T}} \mathrm{Ind}^{\mathbf{G}^F}_{\mathbf{T}^F}\mathrm{reg}_{\mathbf{T}}$. We now use

12.16 LEMMA. *For any rational maximal torus* $\mathbf{T}$ *we have*

$$\mathrm{Ind}^{\mathbf{G}^F}_{\mathbf{T}^F}(\mathrm{reg}_{\mathbf{T}}) = \varepsilon_{\mathbf{T}}\varepsilon_{\mathbf{G}} R^{\mathbf{G}}_{\mathbf{T}}(\mathrm{reg}_{\mathbf{T}})\,\mathrm{St}_{\mathbf{G}}\,.$$

PROOF OF THE LEMMA: This is just the special case of 12.18 below for $R^{\mathbf{G}}_{\mathbf{T}}$. We give this lemma here just to point out that the proof of 12.18 in this special case needs only 12.9 above (and not 12.17). ■

Applying this lemma, we get $\mathrm{reg}_{\mathbf{G}} = |\mathcal{T}|^{-1}\,\mathrm{St}_{\mathbf{G}}\sum_{\mathbf{T}\in\mathcal{T}}\varepsilon_{\mathbf{T}}\varepsilon_{\mathbf{G}} R^{\mathbf{G}}_{\mathbf{T}}(\mathrm{reg}_{\mathbf{T}})$ and the right-hand side above is equal to $|\mathcal{T}|^{-1}\,\mathrm{St}_{\mathbf{G}}\,|\mathbf{G}^F|_p\,\mathrm{reg}_{\mathbf{G}}$ by 12.14. Taking the value at 1 of both sides, we get the result, as $\mathrm{St}_{\mathbf{G}}(1) = |\mathbf{G}^F|_p$. ■

We will now give, using 12.9, the analogue for $R^{\mathbf{G}}_{\mathbf{L}}$ of 12.9 and 12.16. These properties could be proved directly, using the same arguments as for 12.9 and 12.16, if we knew the validity of the Mackey formula in general, and thus the truth of 8.11 for a general Lusztig functor.

12.17 PROPOSITION. *Let* **L** *be a Levi subgroup of* **G**, *and let* $\varphi \in \mathrm{Irr}(\mathbf{L}^F)$; *then*

$$\dim(R_{\mathbf{L}}^{\mathbf{G}}\varphi) = \varepsilon_{\mathbf{G}}\varepsilon_{\mathbf{L}}|\mathbf{G}^F/\mathbf{L}^F|_{p'}\dim(\varphi).$$

PROOF: We have $\varphi(1) = \langle \varphi, \mathrm{reg}_{\mathbf{L}} \rangle_{\mathbf{L}^F}$ and similarly

$$(R_{\mathbf{L}}^{\mathbf{G}}\varphi)(1) = \langle R_{\mathbf{L}}^{\mathbf{G}}\varphi, \mathrm{reg}_{\mathbf{G}} \rangle_{\mathbf{G}^F} = \langle R_{\mathbf{L}}^{\mathbf{G}}\varphi, |\mathbf{G}^F|_p^{-1} \sum_{\mathbf{T}\in\mathcal{T}} \varepsilon_{\mathbf{G}}\varepsilon_{\mathbf{T}} R_{\mathbf{T}}^{\mathbf{G}}(\mathrm{reg}_{\mathbf{T}}) \rangle_{\mathbf{G}^F}$$

(the last equality by 12.14). We now use adjunction to transform the last term above, then apply to it the Mackey formula 11.12, and then again take adjoints; we get

$$\begin{aligned}(R_{\mathbf{L}}^{\mathbf{G}}\varphi)(1) &= |\mathbf{G}^F|_p^{-1} \sum_{\mathbf{T}\in\mathcal{T}} \varepsilon_{\mathbf{T}}\varepsilon_{\mathbf{G}} \langle {}^*R_{\mathbf{T}}^{\mathbf{G}} \circ R_{\mathbf{L}}^{\mathbf{G}}\varphi, \mathrm{reg}_{\mathbf{T}} \rangle_{\mathbf{T}^F} \\ &= |\mathbf{G}^F|_p^{-1} \sum_{\mathbf{T}\in\mathcal{T}} \varepsilon_{\mathbf{T}}\varepsilon_{\mathbf{G}} \langle \sum_{\mathbf{L}^F\backslash\{x\in\mathbf{G}^F \mid {}^x\mathbf{T}\subset\mathbf{L}\}} \mathrm{ad}\, x^{-1} \circ {}^*R_{{}^x\mathbf{T}}^{\mathbf{L}}\varphi, \mathrm{reg}_{\mathbf{T}} \rangle_{\mathbf{T}^F} \\ &= |\mathbf{G}^F|_p^{-1} \sum_{\mathbf{T}\in\mathcal{T}} \varepsilon_{\mathbf{T}}\varepsilon_{\mathbf{G}} \sum_{\mathbf{L}^F\backslash\{x\in\mathbf{G}^F \mid {}^x\mathbf{T}\subset\mathbf{L}\}} \langle \varphi, R_{{}^x\mathbf{T}}^{\mathbf{L}}\, {}^x\mathrm{reg}_{\mathbf{T}} \rangle_{\mathbf{L}^F}.\end{aligned}$$

In the last expression we may take as a new variable ${}^x\mathbf{T}$ which is equivalent to summing over all rational maximal tori of **L**, if we multiply the expression by $|\mathbf{G}^F|/|\mathbf{L}^F|$. We get $|\mathbf{G}^F|_{p'}|\mathbf{L}^F|^{-1}\varepsilon_{\mathbf{G}} \sum_{\mathbf{T}\subset\mathbf{L}} \varepsilon_{\mathbf{T}} \langle \varphi, R_{\mathbf{T}}^{\mathbf{L}}\,\mathrm{reg}_{\mathbf{T}} \rangle_{\mathbf{L}^F}$, which is equal by 12.14 to $\varepsilon_{\mathbf{G}}\varepsilon_{\mathbf{L}}|\mathbf{G}^F|_{p'}/|\mathbf{L}^F|_{p'}\langle \varphi, \mathrm{reg}_{\mathbf{L}} \rangle_{\mathbf{L}^F}$, whence the result. ∎

12.18 COROLLARY. *For any* $\varphi \in \mathrm{Irr}(\mathbf{L}^F)$ *(resp.* $\psi \in \mathrm{Irr}(\mathbf{G}^F)$*) we have*

(i) $$(R_{\mathbf{L}}^{\mathbf{G}}\varphi).\varepsilon_{\mathbf{G}}\,\mathrm{St}_{\mathbf{G}} = \mathrm{Ind}_{\mathbf{L}^F}^{\mathbf{G}^F}(\varphi.\varepsilon_{\mathbf{L}}\,\mathrm{St}_{\mathbf{L}})$$

and

(ii) $${}^*R_{\mathbf{L}}^{\mathbf{G}}(\psi.\varepsilon_{\mathbf{G}}\,\mathrm{St}_{\mathbf{G}}) = \varepsilon_{\mathbf{L}}\,\mathrm{St}_{\mathbf{L}}\,.\,\mathrm{Res}_{\mathbf{L}^F}^{\mathbf{G}^F}\psi.$$

PROOF: It is enough to prove (ii) as (i) is its adjoint. Applying 12.2, and using the value of $\mathrm{St}_{\mathbf{L}}$, the truth of (ii) is equivalent to the fact that for any $\psi \in \mathrm{Irr}(\mathbf{G}^F)$ and any semi-simple element $s \in \mathbf{L}^F$ we have

$$\varepsilon_{C^\circ_{\mathbf{G}}(s)}|C^\circ_{\mathbf{G}}(s)^F|_{p'}^{-1}|C^\circ_{\mathbf{L}}(s)^F|\psi(s)Q^{C^\circ_{\mathbf{G}}(s)}_{C^\circ_{\mathbf{L}}(s)}(1,v^{-1}) = \begin{cases} 0 & \text{if } v \neq 1, \\ \varepsilon_{C^\circ_{\mathbf{L}}(s)}|C^\circ_{\mathbf{L}}(s)^F|_p\psi(s) & \text{if } v = 1 \end{cases} \qquad (*)$$

(we have used the fact that the Steinberg character vanishes outside semi-simple elements).

On the other hand 12.17 gives that for any $\varphi \in \mathrm{Irr}(\mathbf{L}^F)$ we have

$$\langle R_{\mathbf{L}}^{\mathbf{G}}\varphi, \mathrm{reg}_{\mathbf{G}} \rangle_{\mathbf{G}^F} = \varepsilon_{\mathbf{G}}\varepsilon_{\mathbf{L}}|\mathbf{G}^F/\mathbf{L}^F|_{p'}\langle \varphi, \mathrm{reg}_{\mathbf{L}} \rangle_{\mathbf{L}^F}$$

whence

$$ {}^*R_{\mathbf{L}}^{\mathbf{G}}\,\mathrm{reg}_{\mathbf{G}} = \varepsilon_{\mathbf{G}}\varepsilon_{\mathbf{L}}|\mathbf{G}^F/\mathbf{L}^F|_{p'}\,\mathrm{reg}_{\mathbf{L}}\,. $$

If in the equality above we apply the character formula 12.2 to ${}^*R_{\mathbf{L}}^{\mathbf{G}}\,\mathrm{reg}_{\mathbf{G}}$, and compute both sides at some given unipotent element v of $\mathbf{L}^F$, we get

$$ Q_{\mathbf{L}}^{\mathbf{G}}(1, v^{-1}) = \begin{cases} 0 & \text{if } v \neq 1, \\ \varepsilon_{\mathbf{G}}\varepsilon_{\mathbf{L}}|\mathbf{G}^F/\mathbf{L}^F|_{p'} & \text{if } v = 1. \end{cases} \qquad (**) $$

The equality (*) we want to prove results immediately from (**) applied with $\mathbf{G}$ replaced by $C^\circ_{\mathbf{G}}(s)$ and $\mathbf{L}$ replaced by $C^\circ_{\mathbf{L}}(s)$. ■

We end this chapter with a proof that characteristic functions of semi-simple classes are uniform. We use "normalized" characteristic functions.

12.19 NOTATION. *Given a finite group H and some $x \in H$, we denote by π_x^H the function whose value is $|C_H(x)|$ on the H-conjugacy class of x and 0 on other elements of H.*

With this notation, we have

12.20 PROPOSITION. *Let s be a semi-simple element of $\mathbf{G}^F$; then*

$$ \begin{aligned} \pi_s^{\mathbf{G}^F} &= |W^\circ(s)|^{-1} \sum_{w \in W^\circ(s)} \dim(R_{\mathbf{T}_w}^{C^\circ_{\mathbf{G}}(s)}(\mathrm{Id}_{\mathbf{T}_w}))R_{\mathbf{T}_w}^{\mathbf{G}}(\pi_s^{\mathbf{T}_w^F}) \\ &= \varepsilon_{C^\circ_{\mathbf{G}}(s)}|C^\circ_{\mathbf{G}}(s)^F|_p^{-1} \sum_{\substack{\mathbf{T} \in \mathcal{T} \\ \mathbf{T} \ni s}} \varepsilon_{\mathbf{T}} R_{\mathbf{T}}^{\mathbf{G}}(\pi_s^{\mathbf{T}^F}), \end{aligned} $$

where in the first sum $W^\circ(s)$ is as in 2.4, and $\mathbf{T}_w$ is of type w with respect to some fixed torus of $C^\circ_{\mathbf{G}}(s)$.

PROOF: Let $\gamma_p^s \in \mathcal{C}(\mathbf{G}^F)_{p'}$ be the function with value $|C^\circ_{\mathbf{G}}(s)^F|_{p'}$ on elements whose semi-simple part is conjugate to s (the "p'-section" of s) and value 0 elsewhere. Using 9.3 we get

$$ \pi_s^{\mathbf{G}^F} = \varepsilon_{\mathbf{G}}\varepsilon_{C^\circ_{\mathbf{G}}(s)}|C_{\mathbf{G}}(s)^F/C^\circ_{\mathbf{G}}(s)^F|\,\mathrm{St}_{\mathbf{G}}\,\gamma_p^s, $$

which is equal to

$$ |\mathbf{G}^F|^{-1} \sum_{\mathbf{T} \in \mathcal{T}} |\mathbf{T}^F|\varepsilon_{\mathbf{G}}\varepsilon_{\mathbf{T}} R_{\mathbf{T}}^{\mathbf{G}}(\mathrm{Id}_{\mathbf{T}})\gamma_p^s = |\mathbf{G}^F|^{-1} \sum_{\mathbf{T}} |\mathbf{T}^F|\varepsilon_{\mathbf{G}}\varepsilon_{\mathbf{T}} R_{\mathbf{T}}^{\mathbf{G}}(\mathrm{Res}_{\mathbf{T}^F}^{\mathbf{G}^F}\gamma_p^s), $$

(using the formula for $\mathrm{St}_{\mathbf{G}}$ in the proof of 12.14 and 12.6). But the value of $\mathrm{Res}_{\mathbf{T}^F}^{\mathbf{G}^F}\gamma_p^s$ is $|C^\circ_{\mathbf{G}}(s)^F|_{p'}$ on the intersection of the class of s with $\mathbf{T}^F$ and 0 elsewhere, so this function is easily seen to be equal to

$$ |C^\circ_{\mathbf{G}}(s)^F|_{p'}|C_{\mathbf{G}}(s)^F|^{-1} \sum_{\{g \in \mathbf{G}^F \mid {}^g s \in \mathbf{T}\}} |\mathbf{T}^F|^{-1}\pi_{{}^g s}^{\mathbf{T}^F}\,. $$

Using this value we get

$$\begin{aligned}
\pi_s^{\mathbf{G}^F} &= \varepsilon_{C^\circ_{\mathbf{G}}(s)}|\mathbf{G}^F|^{-1}|C^\circ_{\mathbf{G}}(s)^F|_p^{-1}\sum_{\mathbf{T}\in\mathcal{T}}\varepsilon_{\mathbf{T}}\sum_{\{g\in\mathbf{G}^F\mid {}^gs\in\mathbf{T}\}} R_{\mathbf{T}}^{\mathbf{G}}(\pi_{{}^gs}^{\mathbf{T}^F})\\
&= \varepsilon_{C^\circ_{\mathbf{G}}(s)}|\mathbf{G}^F|^{-1}|C^\circ_{\mathbf{G}}(s)^F|_p^{-1}\sum_{g\in\mathbf{G}^F}\sum_{\mathbf{T}\ni {}^gs}\varepsilon_{\mathbf{T}}R_{\mathbf{T}}^{\mathbf{G}}(\pi_{{}^gs}^{\mathbf{T}^F})\\
&= \varepsilon_{C^\circ_{\mathbf{G}}(s)}|\mathbf{G}^F|^{-1}|C^\circ_{\mathbf{G}}(s)^F|_p^{-1}\sum_{g\in\mathbf{G}^F}\sum_{{}^{g^{-1}}\mathbf{T}\ni s}\varepsilon_{{}^{g^{-1}}\mathbf{T}}R_{{}^{g^{-1}}\mathbf{T}}^{\mathbf{G}}(\pi_s^{{}^{g^{-1}}\mathbf{T}^F})\\
&= \varepsilon_{C^\circ_{\mathbf{G}}(s)}|\mathbf{G}^F|^{-1}|C^\circ_{\mathbf{G}}(s)^F|_p^{-1}\sum_{g\in\mathbf{G}^F}\sum_{\mathbf{T}\ni s}\varepsilon_{\mathbf{T}}R_{\mathbf{T}}^{\mathbf{G}}(\pi_s^{\mathbf{T}^F})\\
&= \varepsilon_{C^\circ_{\mathbf{G}}(s)}|C^\circ_{\mathbf{G}}(s)^F|_p^{-1}\sum_{\mathbf{T}\ni s}\varepsilon_{\mathbf{T}}R_{\mathbf{T}}^{\mathbf{G}}(\pi_s^{\mathbf{T}^F})
\end{aligned}$$

(exchanging sums, then using the fact that $R_{\mathbf{T}}^{\mathbf{G}}(\pi_{{}^gs}^{\mathbf{T}^F})$ is a central function on $\mathbf{G}^F$, then taking ${}^{g^{-1}}\mathbf{T}$ as new variable). This gives the second formula of the statement. The first one results from a straightforward computation. ■

Note that $D_{\mathbf{G}}(\gamma_p^s) = D_{\mathbf{G}}(\mathrm{Id}_{\mathbf{G}}.\gamma_p^s) = D_{\mathbf{G}}(\mathrm{Id}_{\mathbf{G}}).\gamma_p^s = \mathrm{St}_{\mathbf{G}}.\gamma_p^s$ so the beginning of the preceding proof gives a formula for $D_{\mathbf{G}}(\gamma_p^s)$ which generalizes 9.4.

12.21 EXERCISE. Show that any function in $\mathcal{C}(\mathbf{G}^F)_{p'}$ is uniform.

HINT: Use 12.6 and 12.13. ■

12.22 EXERCISE. Let $x = su$ be the Jordan decomposition of some element $x \in \mathbf{G}^F$, and let $\mathbf{L}$ be a rational Levi subgroup of $\mathbf{G}$ containing $C^\circ_{\mathbf{G}}(s)$; prove that

$$R_{\mathbf{L}}^{\mathbf{G}}(\pi_x^{\mathbf{L}^F}) = \pi_x^{\mathbf{G}^F}.$$

HINT: Use 12.5 to compute $({}^*R_{\mathbf{L}}^{\mathbf{G}}\psi)(x)$, for any $\psi \in \mathrm{Irr}\,\mathbf{G}^F$. ■

References

The character formula 12.2 for the functor $R_{\mathbf{T}}^{\mathbf{G}}$ is in [DL1, 4.2]. The generalization to $R_{\mathbf{L}}^{\mathbf{G}}$ was first published in [F. DIGNE and J. MICHEL Foncteur de Lusztig et fonctions de Green généralisées, *CRAS.*, **297** (sept. 1983), 89–92]; we have since learned that it was already known by Deligne at the time of [DL1]. The number of maximal tori was given by Steinberg in [St1, 14.14] using a completely different proof. We follow here the arguments of [DM2].

13. GEOMETRIC CONJUGACY AND LUSZTIG SERIES

We will now introduce Lusztig's classification of irreducible characters of $\mathbf{G}^F$. In this classification, a character is parametrized by a pair: a rational semi-simple class (s) of the dual group $\mathbf{G}^*$ of $\mathbf{G}$, and a "unipotent" irreducible character χ of the centralizer $C_{\mathbf{G}^{*F^*}}(s)$.

We have all the tools to deal with the ingredient (s) of the classification, but will be able to prove very little about the ingredient χ here. We start with an immediate corollary of 12.14.

13.1 PROPOSITION. *For any $\chi \in \operatorname{Irr}(\mathbf{G}^F)$, there exists a rational maximal torus $\mathbf{T}$ and $\theta \in \operatorname{Irr}(\mathbf{T}^F)$ such that $\langle \chi, R_{\mathbf{T}}^{\mathbf{G}}(\theta) \rangle_{\mathbf{G}^F} \neq 0$.*

We note also that 11.15 (i) implies that $R_{\mathbf{T}}^{\mathbf{G}}(\theta)$ is irreducible if and only if no non-trivial element of $W(\mathbf{T})^F$ stabilizes θ. Furthermore, by 12.9, in that case $\varepsilon_{\mathbf{T}}\varepsilon_{\mathbf{G}} R_{\mathbf{T}}^{\mathbf{G}}(\theta)$ is a true character.

We will now give a condition for two Deligne-Lusztig characters to have a common irreducible component (using the "norm" on a torus, see 11.9). By 11.15 (i), if the pairs $(\mathbf{T}, \theta)$ and $(\mathbf{T}', \theta')$ are not $\mathbf{G}^F$-conjugate then $R_{\mathbf{T}}^{\mathbf{G}}(\theta)$ and $R_{\mathbf{T}}^{\mathbf{G}}(\theta')$ are orthogonal to each other but they may have a common constituent as they are virtual characters.

13.2 DEFINITION. *Let $\mathbf{T}$ and $\mathbf{T}'$ be two rational maximal tori, and let θ and θ' be characters respectively of $\mathbf{T}^F$ and $\mathbf{T}'^F$. We say that the pairs $(\mathbf{T}, \theta)$ and $(\mathbf{T}', \theta')$ are* **geometrically conjugate** *if there exists $g \in \mathbf{G}$ such that $\mathbf{T} = {}^g\mathbf{T}'$ and such that for any n such that $g \in \mathbf{G}^{F^n}$ we have $\theta \circ N_{F^n/F} = \theta' \circ N_{F^n/F} \circ \operatorname{ad} g$.*

13.3 PROPOSITION. *Let $\mathbf{T}$ and $\mathbf{T}'$ be two rational maximal tori, and let $\mathbf{U}$ (resp. $\mathbf{U}'$) be the unipotent radical of some Borel subgroup containing $\mathbf{T}$ (resp. $\mathbf{T}'$). Let θ and θ' be characters respectively of $\mathbf{T}^F$ and $\mathbf{T}'^F$. Assume that there exist i and j such that the $\mathbf{G}^F$-modules $H_c^i(\mathcal{L}^{-1}(\mathbf{U})) \otimes_{\overline{\mathbb{Q}}_\ell[\mathbf{T}^F]} \theta$ and $H_c^j(\mathcal{L}^{-1}(\mathbf{U}')) \otimes_{\overline{\mathbb{Q}}_\ell[\mathbf{T}'^F]} \theta'$ have a common irreducible constituent; then the pairs $(\mathbf{T}, \theta)$ and $(\mathbf{T}', \theta')$ are geometrically conjugate.*

PROOF: If χ is the common irreducible constituent of the statement, we remark first that χ occurring in $H_c^i(\mathcal{L}^{-1}(\mathbf{U}))\otimes\theta$ is equivalent to $\chi^\vee$ occurring in $\theta^\vee \otimes H^i_c(\mathcal{L}^{-1}(\mathbf{U})^\vee)$ (where, given a left representation χ of a group H, we let $\chi^\vee$ denote the right representation obtained by making elements act through their inverse). As χ occurs in $H^j_c(\mathcal{L}^{-1}(\mathbf{U}')) \otimes_{\overline{\mathbb{Q}}_\ell[\mathbf{T}'^F]} \theta'$, the representation $\theta \otimes \theta'^\vee$ of $\mathbf{T}^F \times \mathbf{T}'^F$ occurs in the module $H^i_c(\mathcal{L}^{-1}(\mathbf{U})^\vee) \otimes_{\overline{\mathbb{Q}}_\ell[\mathbf{G}^F]} H^j_c(\mathcal{L}^{-1}(\mathbf{U}'))$, which with the notation of 11.7 is a submodule of $H^{i+j}_c(\mathbf{Z})$. However, we have:

13.4 LEMMA. *If the $\mathbf{T}^F$-module-${}^w(\mathbf{T}'^F)$ given by $\theta \otimes {}^w\theta'^\vee$ occurs in some cohomology group of $\mathbf{Z}''_w$ (see 11.8) and if $n > 0$ is such that ${}^{F^n}w = w$, then $\theta \circ N_{F^n/F} = \theta' \circ N_{F^n/F} \circ \operatorname{ad} {}^F w^{-1}$.*

PROOF: Using the remark which follows the proof of 11.10, we get the result from 11.10 (iii) applied with $\mathbf{L} = \mathbf{T}$ and $\mathbf{M} = {}^w\mathbf{T}'$, using the fact that $N_{F'^n/F'}({}^{w'}\tau^{-1}) = {}^w N_{F^n/F}({}^{{}^F w^{-1}}\tau^{-1})$. ■

If the hypothesis of the above lemma holds, then the element ${}^F w^{-1}$ is the required element g, and the proposition is proved. As the cohomology of the $\mathbf{T}^F$-variety-${}^w\mathbf{T}'^F$ given by $\mathbf{Z}''_w$ is isomorphic to that of the $\mathbf{T}^F$-variety-$\mathbf{T}'^F$ given by $\mathbf{Z}_w$ (via $\operatorname{ad} w$), the following lemma thus completes the proof of the proposition.

13.5 LEMMA. *If the character $\theta \otimes \theta'^\vee$ does not occur in $H^k_c(\mathbf{Z}_w)$ for any k and w, then $\theta \otimes \theta'^\vee$ does not occur in $H^i_c(\mathbf{Z})$ for any i.*

PROOF: We use the following

13.6 LEMMA.

(i) With the notation of 11.7, if $\mathbf{L} = \mathbf{T}$ and $\mathbf{L}' = \mathbf{T}'$ are tori, and if we let $\mathbf{B}$ and $\mathbf{B}'$ denote the Borel subgroups ${}^{F^{-1}}\mathbf{P}$ and ${}^{F^{-1}}\mathbf{P}'$, then for any $v \in W(\mathbf{T})$ the union $\bigcup_{v' \leq v} \mathbf{Z}_{v'w_1}$ is closed in $\mathbf{Z}$ where $w_1 \in \mathcal{S}(\mathbf{T}, \mathbf{T}')$ is such that ${}^{w_1}\mathbf{B}' = \mathbf{B}$ (and where $v' \leq v$ refers to the Bruhat order, i.e., v' is the product of a sub-sequence extracted from a reduced decomposition of v).

(ii) The connected components of the union $\bigcup_{l(v)=n} \mathbf{Z}_{vw_1}$ are the $\mathbf{Z}_{vw_1}$.

PROOF: Property (i) results from the fact that in the present case $\mathbf{Z}_{vw_1}$ is the inverse image in $\mathbf{Z}$ by the third projection of the subset $\mathbf{B}vw_1\mathbf{B}' = \mathbf{B}v\mathbf{B}w_1$ of $\mathbf{G}$. As the closure of $\mathbf{B}v\mathbf{B}$ in $\mathbf{G}$ is the union $\bigcup_{v' \leq v} \mathbf{B}v'\mathbf{B}$ (see, *e.g.*, [BOREL-TITS, Compléments à l'article: "Groupes Réductifs", *Publications de l'IHES*, **41** (1965), 253–276, §3]), we get (i). Furthermore, (i) shows that the closure of $\mathbf{Z}_{vw_1}$ in $\mathbf{Z}$ does not meet $\mathbf{Z}_{v'w_1}$ if $l(v) = l(v')$, whence (ii). ■

This lemma will allow us to apply repeatedly the long exact sequence 10.7 (i) to $\mathbf{Z} = \bigcup_v \mathbf{Z}_{vw_1}$. We now prove 13.5 by showing by induction on n that the hypothesis of 13.5 implies that $H^i_c(\bigcup_{l(v) \leq n} \mathbf{Z}_{vw_1})_{\theta \otimes \theta'^\vee} = 0$ (where the subscript $\theta \otimes \theta'^{\vee}$ denotes the subspace of the cohomology where $\mathbf{T}^F \times \mathbf{T}'^F$ acts through $\theta \otimes \theta'^{\vee}$). This is true for $n = 0$ by hypothesis. Suppose it for $n - 1$. We have

$$\bigcup_{l(v)\leq n} \mathbf{Z}_{vw_1} = (\bigcup_{l(v)=n} \mathbf{Z}_{vw_1}) \bigcup (\bigcup_{l(v)\leq n-1} \mathbf{Z}_{vw_1}),$$

and this last union is closed since by 13.6 (i) it is a finite union of closed subsets. There is thus a cohomology long exact sequence relating these three unions; this sequence remains exact restricted to the subspaces where $\mathbf{T}^F \times \mathbf{T}'^F$ acts through $\theta \otimes \theta'^{\vee}$. But 13.6 (ii) implies, according to 10.7 (ii), that for any k we have $H^k_c(\bigcup_{l(v)=n} \mathbf{Z}_{vw_1}) = \bigoplus_{l(v)=n} H^k_c(\mathbf{Z}_{vw_1})$, so by assumption $H^k_c(\bigcup_{l(v)=n} \mathbf{Z}_{vw_1})_{\theta\otimes\theta'^{\vee}} = 0$. Since, by the induction hypothesis, for any k we have $H^k_c(\bigcup_{l(v)\leq n-1} \mathbf{Z}_{vw_1})_{\theta\otimes\theta'^{\vee}} = 0$, at each step two out of three terms of the long exact sequence are 0, so the third one $H^k_c(\bigcup_{l(v)\leq n} \mathbf{Z}_{vw_1})_{\theta\otimes\theta'^{\vee}}$ is also. ∎

We will now give some more theory on tori which will allow us to give a nice interpretation of geometric conjugacy (and a justification of the terminology), using the dual of $\mathbf{G}$.

We shall assume chosen once and for all an isomorphism $\overline{\mathbb{F}}_q^{\times} \xrightarrow{\sim} (\mathbb{Q}/\mathbb{Z})_{p'}$ and an embedding $\overline{\mathbb{F}}_q^{\times} \hookrightarrow \overline{\mathbb{Q}}_\ell^{\times}$.

13.7 PROPOSITION. *Let* $\mathbf{T}$ *be a torus defined over* $\mathbb{F}_q$*; we denote by* F *the action of the Frobenius endomorphism on* $X = X(\mathbf{T})$ *(resp.* $Y = Y(\mathbf{T})$*) defined by* $\alpha \mapsto \alpha \circ F$ *(resp.* $\beta \mapsto F \circ \beta$*). Then:*

(i) The sequence

$$0 \to X \xrightarrow{F-1} X \to \mathrm{Irr}(\mathbf{T}^F) \to 1$$

is exact, where the right map is the restriction to $\mathbf{T}^F$ *of characters* (for this to make sense we think of the values of an element of X as being in $\overline{\mathbb{Q}}_\ell$, using the chosen embedding $\overline{\mathbb{F}}_q^{\times} \hookrightarrow \overline{\mathbb{Q}}_\ell^{\times}$).

(ii) The sequence

$$0 \to Y \xrightarrow{F-1} Y \to \mathbf{T}^F \to 1$$

is exact, where the right map is defined by $y \mapsto N_{F^n/F}(y(\zeta))$ *where* n *is such that* $\mathbf{T}$ *is split over* $\mathbb{F}_{q^n}$ *and where* ζ *is the* $(q^n - 1)$*-th root of 1 in* $\overline{\mathbb{F}}_q^{\times}$ *which is the image of* $1/(q^n - 1) \in (\mathbb{Q}/\mathbb{Z})_{p'}$ *by the chosen isomorphism.*

PROOF: The group $\mathbf{T}^F$ is the kernel of $\mathbf{T} \xrightarrow{F-1} \mathbf{T}$, whence an exact sequence

$$1 \to \mathbf{T}^F \to \mathbf{T} \xrightarrow{F-1} \mathbf{T} \to 1.$$

We get (i) by taking the homomorphisms from this sequence to $\mathbb{G}_m$: we have only to check the surjectivity of $X(\mathbf{T}) \to \mathrm{Irr}(\mathbf{T}^F)$. This results from the fact that the algebra of $\mathbf{T}^F$ (viewed as an algebraic group) is the group algebra $\overline{\mathbb{F}}_q[\mathrm{Irr}(\mathbf{T}^F)]$ of the group $\mathrm{Irr}(\mathbf{T}^F)$. To see that, first notice that it is clearly true for a cyclic p'-group and then use that $\mathbf{T}^F$ is a direct product of such groups. As $\mathbf{T}^F$ is a subvariety of $\mathbf{T}$, the corresponding morphism of algebras, $\overline{\mathbb{F}}_q[X(\mathbf{T})] \to \overline{\mathbb{F}}_q[\mathrm{Irr}(\mathbf{T}^F)]$, is surjective, which implies the surjectivity of $X(\mathbf{T}) \to \mathrm{Irr}(\mathbf{T}^F)$ (compare with the argument in 0.5).

We now prove (ii). Let n be the smallest integer such that $\mathbf{T}$ is split over $\mathbb{F}_{q^n}$ (this is the order of τ with the notation of 8.1 and all other n are multiples of that one). For any positive integer k, if $\zeta \in \overline{\mathbb{F}}_q$ is the image of $1/(q^{nk}-1) \in (\mathbb{Q}/\mathbb{Z})_{p'}$, then $\zeta^{(q^{nk}-1)/(q^n-1)}$ is the image of $1/(q^n-1)$ and we have

$$N_{F^n/F}(y(\zeta^{(q^{nk}-1)/(q^n-1)})) = N_{F^n/F}(N_{F^{nk}/F^n}(y(\zeta))) = N_{F^{nk}/F}(y(\zeta)),$$

so the map $Y \to \mathbf{T}^F$ of (ii) is well-defined.

To show the exactness we use the commutative diagram

$$\begin{array}{ccccccccc} 0 & \to & Y & \xrightarrow{F^n-1} & Y & \to & \mathbf{T}^{F^n} & \to & 1 \\ & & \downarrow{\scriptstyle N_{F^n/F}} & & \| & & \downarrow{\scriptstyle N_{F^n/F}} & & \\ 0 & \to & Y & \xrightarrow{F-1} & Y & \to & \mathbf{T}^F & \to & 1 \end{array},$$

where n and the maps from Y to $\mathbf{T}^{F^n}$ and from Y to $\mathbf{T}^F$ are as in the statement, *i.e.*, the top right map sends y to $y(1/(q^n-1))$. We show first the exactness of the top sequence: the injectivity is clear as $F^n - 1 = q^n - 1$ since $\mathbf{T}$ is split over $\mathbb{F}_{q^n}$; the surjectivity is straightforward by 0.20. Moreover if y is in the kernel of the right map then it is trivial on all (q^n-1)-th roots of unity, so is constant on all fibres of the map $x \mapsto x^{q^n-1}$ from $\mathbb{G}_m$ to itself; so it factors through that map, *i.e.*, there exists $y_1 \in Y$ such that $y = (q^n-1)y_1$, *i.e.*, y is in the image of $F^n - 1$. The exactness of the bottom sequence is deduced from that of the top sequence: the surjectivity follows by the surjectivity of $N_{F^n/F} : \mathbf{T}^{F^n} \to \mathbf{T}^F$. The injectivity of $F-1$ is clear as F is the transpose of the endomorphism $q\tau^{-1}$ of $X(\mathbf{T})$ (see formula before 8.1) so has no eigenvalue 1. The image of $F-1$ is obviously in the kernel of $Y \to \mathbf{T}^F$. Take now y in the kernel of the right map, *i.e.*, such that $N_{F^n/F}(y(1/(q^n-1))) = 1$. This means that $N_{F^n/F}(y)$ is in the kernel of $Y \to \mathbf{T}^{F^n}$, so is in $(F^n-1)Y = N_{F^n/F}((F-1)Y)$. As $N_{F^n/F} : Y \to Y$ is injective, since $(F-1) \circ N_{F^n/F} = F^n - 1$ is injective, we get $y \in (F-1)Y$. ∎

We note that using (ii) above any character of $\mathbf{T}^F$ gives a character of $Y(\mathbf{T})$. We may now reinterpret the notion of geometric conjugacy.

13.8 PROPOSITION. *Let* $\mathbf{T}$ *and* $\mathbf{T}'$ *be two rational maximal tori and let* $\theta \in \operatorname{Irr}(\mathbf{T}^F)$ *and* $\theta' \in \operatorname{Irr}(\mathbf{T}'^F)$*; then* $(\mathbf{T}, \theta)$ *and* $(\mathbf{T}', \theta')$ *are geometrically conjugate if and only if there exists an element* $g \in \mathbf{G}$ *which conjugates* $\mathbf{T}$ *to* $\mathbf{T}'$ *and conjugates* θ*, considered as a character of* $Y(\mathbf{T})$*, to* θ'*, considered as a character of* $Y(\mathbf{T}')$*.*

PROOF: The characters θ and $\theta \circ N_{F^n/F}$ are identified with the same character of $Y(\mathbf{T})$ by the construction of 13.7 (ii) (as seen in the proof of that proposition). This gives the result, the element g of the statement being the same as that of 13.2. ■

13.9 COROLLARY. *Let us fix a rational maximal torus* $\mathbf{T}$*; then geometric conjugacy classes of pairs* $(\mathbf{T}', \theta')$ *(where* $\theta' \in \operatorname{Irr}(\mathbf{T}'^F)$*) are in one-to-one correspondence with* F*-stable* $W(\mathbf{T})$*-orbits in* $X(\mathbf{T}) \otimes (\mathbb{Q}/\mathbb{Z})_{p'}$*.*

PROOF: We first remark that all tori are conjugate over $\mathbf{G}$ so we may assume that $\mathbf{T}' = \mathbf{T}$ (so ${}^g\mathbf{T} = \mathbf{T}'$ becomes $g \in W(\mathbf{T})$). The result then comes from the remark that the group of characters of $Y(\mathbf{T})$ is isomorphic to $X(\mathbf{T}) \otimes (\mathbb{Q}/\mathbb{Z})_{p'}$. ■

We now introduce the dual of a reductive group.

13.10 DEFINITION. *Two connected reductive algebraic groups* $\mathbf{G}$ *and* $\mathbf{G}^*$ *are said to be* **dual** *to each other if there exists a maximal torus* $\mathbf{T}$ *of* $\mathbf{G}$ *(resp.* $\mathbf{T}^*$ *of* $\mathbf{G}^*$*) and an isomorphism from* $X(\mathbf{T})$ *to* $Y(\mathbf{T}^*)$ *which sends the roots of* $\mathbf{G}$ *to the coroots of* $\mathbf{T}^*$*. If in addition* $\mathbf{G}$ *and* $\mathbf{G}^*$ *are defined over* $\mathbb{F}_q$ *with respective Frobenius endomorphisms* F *and* F^**, and if* $\mathbf{T}$ *and* $\mathbf{T}^*$ *are rational and the isomorphism above is compatible with the actions of* F *and of* F^**, we say then that the pair* $(\mathbf{G}, F)$ *is dual to the pair* $(\mathbf{G}^*, F^*)$*.*

In particular, two tori $\mathbf{T}$ and $\mathbf{T}^*$ are said to be dual to each other if we have been given an isomorphism $X(\mathbf{T}) \xrightarrow{\sim} Y(\mathbf{T}^*)$. When we talk of groups dual to each other, we will always assume that we have chosen corresponding dual tori; we will say that $(\mathbf{G}, \mathbf{T})$ is dual to $(\mathbf{G}^*, \mathbf{T}^*)$ when we need to specify the tori used. Note that by 0.45 every group has a dual and, given $\mathbf{G}$, the isomorphism class of $\mathbf{G}^*$ is well-defined; however, the duality $X(\mathbf{T}) \xrightarrow{\sim} Y(\mathbf{T}^*)$ is only determined up to a (possibly outer) automorphism of $\mathbf{G}$ stabilizing $(\mathbf{T}, F)$. Using the exact pairing between $X(\mathbf{T})$ and $Y(\mathbf{T})$, being given the above isomorphism $X(\mathbf{T}) \xrightarrow{\sim} Y(\mathbf{T}^*)$ may be seen to be equivalent to being given an isomorphism from $X(\mathbf{T}^*)$ to $Y(\mathbf{T})$ having the same properties. Such an isomorphism makes the $\mathbb{Z}$-dual $Y(\mathbf{T})$ of $X(\mathbf{T})$ isomorphic to $X(\mathbf{T}^*)$; this allows, given any endomorphism φ of $X(\mathbf{T})$, to define the dual endomorphism φ^* of $X(\mathbf{T}^*)$. It is easy to see also that $w \mapsto w^*$ is an anti-isomorphism $W(\mathbf{T}) \to W(\mathbf{T}^*)$, that F^* is the dual of F, and that through the anti-isomorphism the action of F on $W(\mathbf{T})$ is identified with the inverse of the action of F^* on $W(\mathbf{T}^*)$.

EXAMPLES. The group $\mathbf{GL}_n$ is its own dual, and the same is true for the unitary group (which is $\mathbf{GL}_n$ with the Frobenius endomorphism which sends a matrix $(a_{i,j})$ to ${}^t(a_{i,j}^q)^{-1}$). The groups $\mathbf{SL}_n$ and $\mathbf{PGL}_n$ are dual to each other. The group $\mathbf{SO}_{2l}$ is its own dual, with either the standard Frobenius endomorphism or the Frobenius endomorphism of the non-split group (see 15.3). The symplectic group $\mathbf{Sp}_{2n}$ is dual to $\mathbf{SO}_{2n+1}$.

We get as an immediate corollary of 13.7:

13.11 PROPOSITION. *If $(\mathbf{T}^*, F^*)$ is dual to $(\mathbf{T}, F)$ then* $\mathrm{Irr}(\mathbf{T}^F) \simeq \mathbf{T}^{*F^*}$.

The next proposition interprets geometric conjugacy using a group dual to $\mathbf{G}$.

13.12 PROPOSITION. *Assume that $(\mathbf{G}, F)$ and $(\mathbf{G}^*, F^*)$ are dual to each other with corresponding dual tori $\mathbf{T}$ and $\mathbf{T}^*$. Geometric conjugacy classes of pairs $(\mathbf{T}', \theta')$ in $\mathbf{G}$ are in one-to-one correspondence with F^*-stable conjugacy classes of semi-simple elements of $\mathbf{G}^*$.*

PROOF: We apply 13.9. Using the fact that the isomorphism from $X(\mathbf{T})$ to $Y(\mathbf{T}^*)$, is compatible with the action of the Weyl group, we get a bijection between geometric conjugacy classes of pairs $(\mathbf{T}', \theta')$ and F^*-stable $W(\mathbf{T}^*)$-orbits in $Y(\mathbf{T}^*) \otimes (\mathbb{Q}/\mathbb{Z})_{p'}$. But by 0.20 there is an isomorphism from $Y(\mathbf{T}^*) \otimes (\mathbb{Q}/\mathbb{Z})_{p'}$ to $\mathbf{T}^*$ (which depends on the fixed isomorphism $\overline{\mathbb{F}}_q^\times \simeq (\mathbb{Q}/\mathbb{Z})_{p'}$), so geometric conjugacy classes of pairs $(\mathbf{T}', \theta')$ are in one-to-one correspondence with F^*-stable $W(\mathbf{T}^*)$-orbits in $\mathbf{T}^*$. As any semi-simple element is in a maximal torus and all maximal tori are conjugate, any semi-simple class meets $\mathbf{T}^*$. Furthermore, by 0.12 (iv), two elements of $\mathbf{T}^*$ are geometrically conjugate if and only if they are in the same $W(\mathbf{T}^*)$-orbit; and F^*-stable orbits clearly correspond to F^*-stable conjugacy classes, whence the result. ∎

We note that an F^*-stable conjugacy class contains rational elements by the Lang-Steinberg theorem (see 3.12), so F^*-stable geometric conjugacy classes are also the geometric conjugacy classes (*i.e.*, classes under $\mathbf{G}^*$) of rational elements.

After 11.15 we remarked that the $R_{\mathbf{T}}^{\mathbf{G}}(\theta)$ are parametrized by $\mathbf{G}^F$-conjugacy classes of pairs $(\mathbf{T}, \theta)$. Using the dual group, we may give another parametrization

13.13 PROPOSITION. *The $\mathbf{G}^F$-conjugacy classes of pairs $(\mathbf{T}, \theta)$ where $\mathbf{T}$ is a rational maximal torus of $\mathbf{G}$ and $\theta \in \mathrm{Irr}(\mathbf{T}^F)$ are in one-to-one cor-*

respondence with the $\mathbf{G}^{*F^*}$*-conjugacy classes of pairs* $(\mathbf{T}^*, s)$ *where* s *is a semi-simple element of* $\mathbf{G}^{*F^*}$ *and* $\mathbf{T}^*$ *is a rational maximal torus containing* s.

PROOF: By conjugating $\mathbf{T}'$ to a fixed torus $\mathbf{T}$, we see (see remarks after 3.24) that the $\mathbf{G}^F$-conjugacy classes of pairs $(\mathbf{T}', \theta')$ correspond one-to-one to the conjugacy classes under $W(\mathbf{T})$ of pairs (wF, θ) where $\theta \in \mathrm{Irr}(\mathbf{T}^{wF})$. If $\mathbf{T}$ is the torus chosen to put $\mathbf{G}$ and $\mathbf{G}^*$ in duality, these are by 13.11 in one-to-one correspondence with the conjugacy classes under $W(\mathbf{T}^*)$ of pairs (F^*w^*, s) where $s \in \mathbf{T}^{*F^*w^*}$. In the same way as we did in $\mathbf{G}$ in the beginning of the proof, we may conjugate in $\mathbf{G}^*$ arbitrary pairs $(\mathbf{T}'^*, s)$ to such a form. ■

Using this proposition, we will, in what follows, sometimes use the notation $R^{\mathbf{G}}_{\mathbf{T}^*}(s)$ for $R^{\mathbf{G}}_{\mathbf{T}}(\theta)$.

Lusztig's classification of characters is considerably simpler when centralizers of semi-simple elements in $\mathbf{G}^*$ are connected. We will sometimes need to assume this hypothesis in this chapter and the next. We now show that it holds if the centre of $\mathbf{G}$ is connected, and give some other consequences of the connectedness of $Z(\mathbf{G})$ we will need in the next chapter.

13.14 LEMMA. *Let* $\mathbf{G}$ *be a maximal torus of* $\mathbf{G}$ *and* Φ *be the set of roots of* $\mathbf{G}$ *relative to* $\mathbf{T}$*; then*

(i) *The group* $\mathrm{Irr}(Z(\mathbf{G})/Z(\mathbf{G})^\circ)$ *is canonically isomorphic to the torsion group of* $X(\mathbf{T})/{<}\Phi{>}^{\perp\perp}$.
(ii) *If the centre of* $\mathbf{G}$ *is connected, then the centre of a Levi subgroup of* $\mathbf{G}$ *is also connected.*
(iii) *If* $(\mathbf{G}^*, \mathbf{T}^*)$ *is dual to* $(\mathbf{G}, \mathbf{T})$*, then for any* $s \in \mathbf{T}^*$ *the group* $W(s)/W^\circ(s)$ *(see notation of 2.4) is isomorphic to a subgroup of* $\mathrm{Irr}(Z(\mathbf{G})/Z(\mathbf{G})^\circ)$.

PROOF: Since an element $x \in X(Z(\mathbf{G}))$ is of finite order if and only if the group $Z(\mathbf{G})/\ker x$ is finite, which is equivalent to $\ker x$ containing $Z(\mathbf{G})^\circ$, we see that the group $\mathrm{Irr}(Z(\mathbf{G})/Z(\mathbf{G})^\circ)$ is the torsion group of $X(Z(\mathbf{G}))$. As we have $Z(\mathbf{G}) = {<}\Phi{>}^\perp$, it follows that $X(Z(\mathbf{G})) = X(\mathbf{T})/{<}\Phi{>}^{\perp\perp}$, whence (i).

We see that $Z(\mathbf{G})$ is connected if and only if $X(\mathbf{T})/{<}\Phi{>}^{\perp\perp}$ has no torsion, which is equivalent to $X(\mathbf{T})/{<}\Phi{>}$ having no p'-torsion by 0.24; but the subgroup ${<}\Phi_{\mathbf{L}}{>}$ of ${<}\Phi{>}$ spanned by the roots of a Levi subgroup $\mathbf{L}$ containing $\mathbf{T}$ is a direct summand of ${<}\Phi{>}$, so if $X(\mathbf{T})/{<}\Phi{>}$ has no p'-torsion then $X(\mathbf{T})/{<}\Phi_{\mathbf{L}}{>}$ has none either, so $Z(\mathbf{L})$ is connected, whence (ii).

By 0.20 an element $s \in \mathbf{T}^*$ can be identified with an element of $Y(\mathbf{T}^*) \otimes (\mathbb{Q}/\mathbb{Z})_{p'}$, *i.e.*, of $X(\mathbf{T}) \otimes (\mathbb{Q}/\mathbb{Z})_{p'}$. Let us write such an element x/m with $x \in X(\mathbf{T})$ and $m \in \mathbb{Z}$ relatively prime to p. We have

$$W(s) = \{\, w \in W \mid w(x) - x \in mX(\mathbf{T}) \,\}.$$

Consider the map

$$W(s) \to X(\mathbf{T})/<\Phi>$$
$$w \mapsto \frac{w(x)-x}{m} \quad (\text{mod } <\Phi>);$$

it is a group morphism because, as W acts trivially on $X(\mathbf{T})/<\Phi>$, we have

$$\frac{ww'(x)-w(x)}{m} \equiv \frac{w'(x)-x}{m} \quad (\text{mod } <\Phi>)$$

which is equivalent to

$$\frac{ww'(x)-x}{m} \equiv \frac{w'(x)-x}{m} + \frac{w(x)-x}{m} \quad (\text{mod } <\Phi>).$$

The kernel of this morphism is $\{\, w \in W \mid w(x) - x \in m<\Phi> \}$ which is a reflection group by [Bbk, VI, ex. 1 of §2], so is equal to $W^\circ(s)$ by 2.4. Whence we get an imbedding of $W(s)/W^\circ(s)$ into $X(\mathbf{T})/<\Phi>$. The image is a p'-torsion group since for any w we have $w(x) - x \in \Phi$ (this is true for a reflection, whence also for an arbitrary product of reflections), so $m((w(x)-x)/m) \in \Phi$, which means that the exponent of the image divides m. We then get (iii) by (i) and 0.24. ■

13.15 REMARKS.

(i) Note that the above proof shows that, for any semi-simple element s of a connected algebraic group $\mathbf{G}$, the exponent of $C_{\mathbf{G}}(s)/C^\circ_{\mathbf{G}}(s)$ divides the order of s.

(ii) A consequence of 13.14 (iii) is that, if the centre of $\mathbf{G}$ is connected, the centralizer of any semi-simple element of $\mathbf{G}^*$ is connected.

The next definition is a first step towards the classification of characters of $\mathbf{G}^F$.

13.16 DEFINITION. *A* **Lusztig series** $\mathcal{E}(\mathbf{G}^F, (s))$ *associated to the geometric conjugacy class* (s) *of a semi-simple element* $s \in \mathbf{G}^{*F^*}$ *is the set of irreducible characters of* $\mathbf{G}^F$ *which occur in some* $R^{\mathbf{G}}_{\mathbf{T}}(\theta)$, *where* $(\mathbf{T}, \theta)$ *is of the geometric conjugacy class associated by 13.12 to* (s).

Using the notation after 13.13, we may also define $\mathcal{E}(\mathbf{G}^F, (s))$ as the set of constituents of the $R^{\mathbf{G}}_{\mathbf{T}^*}(s')$ for s' geometrically conjugate to s. The series $\mathcal{E}(\mathbf{G}^F, (s))$ are sometimes called "geometric series"; we will also (see 14.41) consider "rational series", where the rational class of s is fixed. The two notions coincide, by 3.25, when $C_{\mathbf{G}^*}(s)$ is connected.

13.17 PROPOSITION. *Lusztig series associated to various geometric conjugacy classes of semi-simple elements of* $\mathbf{G}^{*F^*}$ *form a partition of* $\mathrm{Irr}(\mathbf{G}^F)$.

PROOF: By 13.3 and 13.12 two Deligne-Lusztig characters in different series have no common constituent, and by 13.1 any irreducible character is in some series. ∎

13.18 REMARK. If $\mathbf{TU}$ is the Levi decomposition of some Borel subgroup containing $\mathbf{T}$, by 13.3 any irreducible character of $\mathbf{G}$ which occurs in some $H^i_c(\mathcal{L}^{-1}(\mathbf{U}))\otimes_{\overline{\mathbb{Q}}_\ell[\mathbf{T}^F]}\theta$ is in the series $\mathcal{E}(\mathbf{G}^F,(s))$, where (s) is the semi-simple conjugacy class of $\mathbf{G}^*$ corresponding to the geometric class of $(\mathbf{T},\theta)$.

A particularly important series, which is a kind of "prototype" for the other ones, is the series associated to the identity element of $\mathbf{G}^*$.

13.19 DEFINITION. *The elements of* $\mathcal{E}(\mathbf{G}^F,(1))$ *(i.e., the irreducible components of the* $R^{\mathbf{G}}_{\mathbf{T}}(\mathrm{Id}_{\mathbf{T}})$*) are called* **unipotent characters**.

The next statement shows that the set of unipotent characters depends only on the isomorphism type of the root system of $\mathbf{G}$.

13.20 PROPOSITION. *Let* $\mathbf{G}$ *and* $\mathbf{G}_1$ *be two reductive groups defined over* $\mathbb{F}_q$*, and let* $f:\mathbf{G}\to\mathbf{G}_1$ *be a morphism of algebraic groups with a central kernel, defined over* $\mathbb{F}_q$ *and such that* $f(\mathbf{G})$ *contains the derived group* $\mathbf{G}_1'$*; then the unipotent characters of* $\mathbf{G}^F$ *are the* $\chi\circ f$*, where* χ *runs over the unipotent characters of* $\mathbf{G}_1^F$.

PROOF: We first remark that in the situation of the proposition we have $\mathbf{G}_1=Z(\mathbf{G}_1)^\circ.f(\mathbf{G})$ by 0.40. The Borel subgroups, the maximal tori and the parabolic subgroups of $f(\mathbf{G})$ are thus the intersections of those of $\mathbf{G}_1$ with $f(\mathbf{G})$. On the other hand, f induces a bijection from the Borel subgroups, the maximal tori and the parabolic subgroups of $\mathbf{G}$ to those of $f(\mathbf{G})$, so there is a bijection from the sets of Borel subgroups, maximal tori and parabolic subgroups of $\mathbf{G}$ to the corresponding sets for $\mathbf{G}_1$.

The unipotent characters of $\mathbf{G}^F$ are those which occur in some cohomology space of some variety $\mathcal{L}_{\mathbf{G}}^{-1}(\mathbf{U})/\mathbf{T}^F$ where $\mathbf{T}$ runs over a set of representatives of classes of rational maximal tori, and where $\mathbf{U}$ is the unipotent radical of some Borel subgroup containing $\mathbf{T}$. The group $\mathbf{U}$ is isomorphic to its image by f which we will denote by $\mathbf{U}_1$. The Borel subgroup of $\mathbf{G}_1$ which corresponds to $\mathbf{TU}$ by the above bijection is $\mathbf{T}_1\mathbf{U}_1$, where $\mathbf{T}_1=f(\mathbf{T}).Z(\mathbf{G}_1)^\circ$. It is clear that f induces a morphism from $\mathcal{L}_{\mathbf{G}}^{-1}(\mathbf{U})/\mathbf{T}^F$ to $\mathcal{L}_{\mathbf{G}_1}^{-1}(\mathbf{U}_1)/\mathbf{T}_1^F$ which is compatible with the actions of $\mathbf{G}^F$ and $\mathbf{G}_1^F$; let us show that it is an isomorphism by exhibiting its inverse. Let $x_1\in\mathbf{G}_1$ be such that $x_1^{-1}{}^Fx_1\in\mathbf{U}_1$;

then any inverse image x of x_1 in $\mathbf{G}$ satisfies $x^{-1}{}^F x \in \mathbf{U}z$ for some $z \in Z(\mathbf{G})$ (depending on x), which may be written $x^{-1}z^F x \in \mathbf{U}$. Let $t \in \mathbf{T}$ be such that $t^{-1}{}^F t = z$; then the element xt is in $\mathcal{L}^{-1}(\mathbf{U})$ and, as $f(t)$ is in $\mathbf{T}_1^F$, the image of $xt\mathbf{T}^F$ in $\mathcal{L}_{\mathbf{G}_1}^{-1}(\mathbf{U}_1)/\mathbf{T}_1^F$ is equal to the image of $x\mathbf{T}^F$, *i.e.*, to $x_1\mathbf{T}_1^F$. As the class $xt\mathbf{T}^F$ is clearly well-defined, we have constructed the required inverse morphism.

We have thus proved that $R_{\mathbf{T}}^{\mathbf{G}}(\mathrm{Id}_{\mathbf{T}}) = R_{\mathbf{T}_1}^{\mathbf{G}_1}(\mathrm{Id}_{\mathbf{T}_1})\circ f$. Moreover as $\mathbf{G}^F$-classes of rational maximal tori are in one-to-one correspondence with $\mathbf{G}_1^F$-classes (they are parametrized by the F-classes of the Weyl group), the "restriction through f" defines a bijection from the set of Deligne-Lusztig characters $R_{\mathbf{T}}^{\mathbf{G}}(\mathrm{Id}_{\mathbf{T}})$ of $\mathbf{G}^F$ onto the similar set for $\mathbf{G}_1^F$. It remains to see that the restriction through f maps the irreducible components of one to those of the other.

Let $\mathbf{T}$ be a rational maximal torus of $\mathbf{G}$ and let $\mathbf{T}_1 = f(\mathbf{T}).Z(\mathbf{G}_1)^\circ$ be the corresponding rational maximal torus of $\mathbf{G}_1$. We have $\mathbf{G}_1^F = \mathbf{T}_1^F.f(\mathbf{G}^F)$. Indeed, any $y \in \mathbf{G}_1$ can be written $y = zf(x)$ with $z \in Z(\mathbf{G}_1)^\circ$ and $x \in \mathbf{G}$, and y is rational if and only if $f(x^F x^{-1}) = z^{-1}{}^F z$. This element is in $Z(\mathbf{G}_1)^\circ \cap f(\mathbf{G}) \subset Z(f(\mathbf{G}))$ and this last group is in $f(\mathbf{T})$ which is a maximal torus of the reductive group $f(\mathbf{G})$ (see the beginning of the present proof). So $x^F x^{-1} \in \mathbf{T}$, since the kernel of f is in $\mathbf{T}$, and by the Lang-Steinberg theorem applied in the group $\mathbf{T}$ we can find $t \in \mathbf{T}$ such that $tx \in \mathbf{G}^F$. So $y = (zf(t^{-1}))f(tx) \in \mathbf{T}_1^F.f(\mathbf{G}^F)$.

From what we have just proved we see that the quotient group $\mathbf{G}_1^F/f(\mathbf{G}^F)$ is commutative, so we can use the following result from Clifford's theory (see [CuR, 11.4, 11.5]).

13.21 LEMMA. *Let G be a finite group and H be a subgroup of G such that the quotient is abelian. Then if χ and χ' are irreducible characters of G we have*

$$\langle \mathrm{Res}_H^G \chi, \mathrm{Res}_H^G \chi' \rangle_H = \#\{\, \zeta \in \mathrm{Irr}(G/H) \mid \chi\zeta = \chi' \,\}.$$

From this lemma we see that the restriction of a unipotent character χ of $\mathbf{G}_1^F$ to $f(\mathbf{G}^F)$ is irreducible if and only if for any irreducible non-trivial character ζ of $\mathbf{G}_1^F/f(\mathbf{G}^F)$ we have $\chi \neq \chi\zeta$. But $\chi\zeta$ is not unipotent if ζ is not trivial, because if χ is a component of $R_{\mathbf{T}_1}^{\mathbf{G}_1}(\mathrm{Id}_{\mathbf{T}_1})$ then $\chi\zeta$ is a component of $R_{\mathbf{T}_1}^{\mathbf{G}_1}(\mathrm{Res}_{\mathbf{T}_1^F}^{\mathbf{G}_1^F} \zeta)$ and the restriction of ζ to $\mathbf{T}_1^F$ is not trivial, since $\mathbf{G}_1^F = \mathbf{T}_1^F.f(\mathbf{G}^F)$, so cannot be geometrically conjugate to the trivial character, whence the proposition. ∎

There is a similar though weaker result for a general Lusztig functor.

13.22 PROPOSITION. *We assume the same hypotheses as in 13.20, and assume in addition that the kernel of f is connected. Then, if $\mathbf{L}_1$ is an F-stable Levi subgroup of some F-stable parabolic subgroup $\mathbf{P}_1$ of $\mathbf{G}_1$ and if $\pi_1 \in \mathrm{Irr}(\mathbf{L}_1^F)$, we have*

$$R^{\mathbf{G}_1}_{\mathbf{L}_1\subset\mathbf{P}_1}(\pi_1)\circ f = R^{\mathbf{G}}_{f^{-1}(\mathbf{L}_1)\subset f^{-1}(\mathbf{P}_1)}(\pi_1\circ f).$$

PROOF: By the remarks we made at the beginning of the proof of 13.20, the group $f^{-1}(\mathbf{P}_1) = f^{-1}(\mathbf{P}_1 \cap f(\mathbf{G}))$ is a parabolic subgroup of $\mathbf{G}$ which has $f^{-1}(\mathbf{L}_1) = f^{-1}(\mathbf{L}_1 \cap f(\mathbf{G}))$ as Levi subgroup, so the statement makes sense. We put $\mathbf{P} = f^{-1}(\mathbf{P}_1)$ and $\mathbf{L} = f^{-1}(\mathbf{L}_1)$. If $\mathbf{P} = \mathbf{L}\mathbf{U}$ is the Levi decomposition of $\mathbf{P}$, and $\mathbf{P}_1 = \mathbf{L}_1\mathbf{U}_1$ that of $\mathbf{P}_1$, then we have $\mathbf{U}_1 = f(\mathbf{U})$ since $\mathbf{U}_1 \subset \mathbf{G}_1' \subset f(\mathbf{G})$.

To prove the theorem, we will use powers F^n of F and take the "limit on n" using 10.5. We have by 11.2

$$(R^{\mathbf{G}_1}_{\mathbf{L}_1\subset\mathbf{P}_1}(\pi_1)\circ f)(g) = |\mathbf{L}_1^F|^{-1} \sum_{l_1\in\mathbf{L}_1^F} \mathrm{Trace}((f(g),l_1)|H_c^*(\mathcal{L}^{-1}(\mathbf{U}_1)))\pi_1(l_1^{-1})$$

and

$$(R^{\mathbf{G}}_{\mathbf{L}\subset\mathbf{P}}(\pi_1\circ f))(g) = |\mathbf{L}^F|^{-1} \sum_{l\in\mathbf{L}^F} \mathrm{Trace}((g,l)|H_c^*(\mathcal{L}^{-1}(\mathbf{U})))\pi_1(f(l)^{-1}).$$

By 10.5 the right-hand sides of the above equalities are equal if, for any n such that $\mathbf{U}$ (and thus $\mathbf{U}_1$ also) is defined over $\mathbb{F}_{q^n}$, then the number of fixed points

$$|\mathbf{L}_1^F|^{-1} \sum_{l_1\in\mathbf{L}_1^F} \pi_1(l_1^{-1})\#\{x\in\mathbf{G}_1 \mid x^{-1}.{}^Fx \in \mathbf{U}_1 \text{ and } f(g).{}^{F^n}xl_1 = x\} \qquad (1)$$

and

$$|\mathbf{L}^F|^{-1} \sum_{l\in\mathbf{L}^F} \pi_1(f(l^{-1}))\#\{x\in\mathbf{G} \mid x^{-1}.{}^Fx \in \mathbf{U} \text{ and } f(g).{}^{F^n}xl = x\}, \qquad (2)$$

are equal. To prove this for any such n we first transform (1). If $x \in \mathbf{G}_1$ is such that $x^{-1}.{}^Fx \in \mathbf{U}_1$ then there exists $\lambda \in \mathbf{L}_1^F$ such that $x\lambda \in f(\mathbf{G})$. Indeed, since $\mathbf{L}_1 \supset Z(\mathbf{G}_1)^\circ$, there exists $l \in \mathbf{L}_1$ such that $xl \in f(\mathbf{G})$. We have $l^Fl^{-1} \in x^{-1}f(\mathbf{G})^Fx = f(\mathbf{G})x^{-1F}x \subset f(\mathbf{G})\mathbf{U}_1 = f(\mathbf{G})$, so by the Lang-Steinberg theorem applied in $\mathbf{L}\cap f(\mathbf{G})$ there exists $l' \in \mathbf{L}\cap f(\mathbf{G})$ such that $l'.{}^Fl'^{-1} = l.{}^Fl^{-1}$. We get the required element λ by setting $\lambda = l'^{-1}l$. If we introduce $y = x\lambda$, there are $|f(\mathbf{L})^F|$ pairs $(y,\lambda) \in f(\mathbf{G}) \times \mathbf{L}_1^F$ such that $y\lambda^{-1} = x$ so (1) can be rewritten

$$|f(\mathbf{L})^F|^{-1}|\mathbf{L}_1^F|^{-1} \sum_{l_1\in\mathbf{L}_1^F,\lambda\in\mathbf{L}_1^F} \pi_1(l_1^{-1})\#\{y \in f(\mathbf{G}) \mid y^{-1}.{}^F y \in \mathbf{U}_1 \text{ and } f(g)^{F^n} y\lambda^{-1}l_1\lambda = y\}$$

which, summing over $l = \lambda^{-1}l_1\lambda$ and using the fact that $\pi_1(\lambda^{-1}l_1^{-1}\lambda) = \pi_1(l_1^{-1})$, may be further rewritten as

$$|f(\mathbf{L})^F|^{-1} \sum_{l\in\mathbf{L}_1^F} \pi_1(l^{-1})\#\{\, y \in f(\mathbf{G}) \mid y^{-1F}y \in \mathbf{U}_1 \text{ and } f(g).^{F^n}yl = y\,\}. \quad (3)$$

We will now transform (2) using the fact that the map

$$\{(x,l) \in \mathbf{G}\times\mathbf{L}^F \mid x^{-1}.{}^F x \in \mathbf{U} \text{ and } g.^{F^n}xl = x\} \to$$
$$\{(y,l_1) \in f(\mathbf{G}) \times f(\mathbf{L})^F \mid y^{-1F}y \in \mathbf{U}_1 \text{ and } f(g)^{F^n}yl_1 = y\}$$

defined by $(x,l) \mapsto (f(x), f(l))$ is surjective and has fibres of the form $\{(xz,l) \mid z \in (\ker f)^F\}$. Indeed, let (y,l_1) be in the right-hand side set; for any x such that $f(x) = y$ we have $x^{-1}.{}^F x \in \mathbf{U}.\ker f$, so by the Lang-Steinberg theorem in $\ker f$ (assumed connected), there exists an inverse image x of y such that $x^{-1}.{}^F x \in \mathbf{U}$. Let l be any inverse image of l_1 in $\mathbf{L}^F$; it is enough to see that the element $z = x^{-1}g.^{F^n}xl \in \ker f$ is in fact in $(\ker f)^F$ (we will get then an inverse image of (y,l_1) by replacing l by lz^{-1}). But we have, using the fact that z is central to get the first equality,

$$x^{-1F}x.z^{-1F}z = (xz)^{-1}.{}^F(xz) = l^{-1F^n}(x^{-1F}x)l = {}^{l^{-1}F^n}(x^{-1}.{}^F x).$$

As $x^{-1}.{}^F x \in \mathbf{U}$, the equality of the extremal terms above shows that $z^{-1}.{}^F z \in \mathbf{U}$, which implies $z = {}^F z$.

Using the above map, we may rewrite (2) as

$$|\mathbf{L}^F|^{-1}|(\ker f)^F| \sum_{l\in f(\mathbf{L})^F} \pi_1(l^{-1})\#\{\, y \in f(\mathbf{G}) \mid y^{-1F}y \in \mathbf{U}_1 \text{ and } f(g).^{F^n}yl = y\,\}.$$

To finish proving the equality of (2) and (3) we remark that, as $\ker f$ is connected, $f(\mathbf{L})^F \simeq (\mathbf{L}/\ker(f))^F \simeq \mathbf{L}^F/(\ker f)^F$, and on the other hand in (3) the only $l \in \mathbf{L}_1^F$ which give a non-zero contribution are those in $f(\mathbf{L})^F$ because of the condition $f(g).^{F^n}yl = y$. ■

We now give without proof a statement of the main result of Lusztig's classification of characters of finite groups of Lie type (which, for groups with connected centre is the subject of Lusztig's book [L4] and was completed for other groups in [G. LUSZTIG On the representations of reductive groups with disconnected center, *Astérisque,* **168** (1988), 157–166]). In order to do that, we first generalize the definition of Deligne-Lusztig characters to non-connected groups by setting $R^{\mathbf{G}}_{\mathbf{T}^\circ}(\theta) = \mathrm{Ind}^{\mathbf{G}^F}_{\mathbf{G}^{\circ F}}(R^{\mathbf{G}^\circ}_{\mathbf{T}^\circ}(\theta))$, where $\mathbf{T}^\circ$ is a rational maximal torus of $\mathbf{G}^\circ$ (actually this character $R^{\mathbf{G}}_{\mathbf{T}^\circ}(\theta)$ is a sum of some characters $R^{\mathbf{G}}_{\mathbf{T}}(\theta')$ which may be defined using a "quasi-torus" $\mathbf{T} = N_{\mathbf{G}}(\mathbf{T}^\circ, \mathbf{B})$ where $\mathbf{B}$ is a Borel subgroup which contains $\mathbf{T}^\circ$). This definition allows us to extend the definition of unipotent characters to $\mathbf{G}^F$ (keeping the same definition, *i.e.*, $\mathcal{E}(\mathbf{G}^F,(1))$ is defined as the set of components of the $R^{\mathbf{G}}_{\mathbf{T}^\circ}(\mathrm{Id}_{\mathbf{T}^\circ})$). With these definitions we may state

13.23 THEOREM. *Let* $\mathbf{G}$ *a connected reductive group; for any semi-simple element* $s \in \mathbf{G}^{*F^*}$, *there is a bijection from* $\mathcal{E}(\mathbf{G}^F,(s))$ *to* $\mathcal{E}(C_{\mathbf{G}^*}(s)^{F^*},1)$, *and this bijection may be chosen such that, extended by linearity to virtual characters, it sends* $\varepsilon_{\mathbf{G}} R^{\mathbf{G}}_{\mathbf{T}^*}(s)$ *(see remark after 13.13) to* $\varepsilon_{C_{\mathbf{G}^*}(s)^\circ} R^{C_{\mathbf{G}^*}(s)}_{\mathbf{T}^*}(\mathrm{Id}_{\mathbf{T}^*})$ *for any rational maximal torus* $\mathbf{T}^*$ *of* $C_{\mathbf{G}^*}(s)$.

In Lusztig's book [L4], this statement is given when the centre of $\mathbf{G}$ is connected. In that case the group $C_{\mathbf{G}^*}(s)$ is connected for any semi-simple element $s \in \mathbf{G}^*$ (see 13.15 (ii)) and the remarks above about non-connected groups are not needed.

13.24 REMARK. *If* ψ_s *is the bijection of 13.23, then for any* $\chi \in \mathcal{E}(\mathbf{G}^F,(s))$ *we have*

$$\chi(1) = \frac{|\mathbf{G}^F|_{p'}}{|C_{\mathbf{G}^*}(s)^{F^*}|_{p'}} \psi_s(\chi)(1).$$

PROOF: Since the characteristic function of the identity is uniform (see 12.14), χ has the same dimension as its projection $p(\chi)$ on the space of uniform functions. Using 13.13, this projection may be written:

$$p(\chi) = \sum_{(\mathbf{T}^*)} \frac{\langle \chi, R^{\mathbf{G}}_{\mathbf{T}^*}(s) \rangle_{\mathbf{G}^F}}{\langle R^{\mathbf{G}}_{\mathbf{T}^*}(s), R^{\mathbf{G}}_{\mathbf{T}^*}(s) \rangle_{\mathbf{G}^F}} R^{\mathbf{G}}_{\mathbf{T}^*}(s) \tag{1}$$

where $\mathbf{T}^*$ runs over $\mathbf{G}^{*F^*}$-conjugacy classes of tori containing s. An easy computation, using 11.15, shows that this can be rewritten

$$p(\chi) = \sum_{(\mathbf{T}^*)} \frac{\langle \chi, R^{\mathbf{G}}_{\mathbf{T}^*}(s) \rangle_{\mathbf{G}^F}}{\langle R^{C_{\mathbf{G}^*}(s)}_{\mathbf{T}^*}(\mathrm{Id}_{\mathbf{T}^*}), R^{C_{\mathbf{G}^*}(s)}_{\mathbf{T}^*}(\mathrm{Id}_{\mathbf{T}^*}) \rangle_{C_{\mathbf{G}^*}(s)^{F^*}}} R^{\mathbf{G}}_{\mathbf{T}^*}(s),$$

where $\mathbf{T}^*$ runs this time over $C_{\mathbf{G}^*}(s)^{F^*}$-classes of tori of $C_{\mathbf{G}^*}(s)$. Since

$$R^{\mathbf{G}}_{\mathbf{T}^*}(s)(1) = \varepsilon_{\mathbf{G}} \varepsilon_{C_{\mathbf{G}^*}(s)^\circ} \frac{|\mathbf{G}^F|_{p'}}{|C_{\mathbf{G}^*}(s)^{F^*}|_{p'}} R^{C_{\mathbf{G}^*}(s)}_{\mathbf{T}^*}(\mathrm{Id}_{\mathbf{T}^*})(1)$$

and

$$\langle \chi, R_{\mathbf{T}^*}^{\mathbf{G}}(s) \rangle_{\mathbf{G}^F} = \varepsilon_{\mathbf{G}} \varepsilon_{C_{\mathbf{G}^*}(s)^\circ} \langle \psi_s(\chi), R_{\mathbf{T}^*}^{C_{\mathbf{G}^*}(s)}(\mathrm{Id}_{\mathbf{T}^*}) \rangle_{C_{\mathbf{G}^*}(s)^{F^*}},$$

(by 13.23) this gives:

$$\chi(1) = \frac{|\mathbf{G}^F|_{p'}}{|C_{\mathbf{G}^*}(s)^{F^*}|_{p'}} \sum_{(\mathbf{T}^*)} \frac{\langle \psi_s(\chi), R_{\mathbf{T}^*}^{C_{\mathbf{G}^*}(s)}(\mathrm{Id}_{\mathbf{T}^*}) \rangle_{C_{\mathbf{G}^*}(s)^{F^*}}}{\langle R_{\mathbf{T}^*}^{C_{\mathbf{G}^*}(s)}(\mathrm{Id}_{\mathbf{T}^*}), R_{\mathbf{T}^*}^{C_{\mathbf{G}^*}(s)}(\mathrm{Id}_{\mathbf{T}^*}) \rangle_{C_{\mathbf{G}^*}(s)^{F^*}}} R_{\mathbf{T}^*}^{C_{\mathbf{G}^*}(s)}(\mathrm{Id}_{\mathbf{T}^*})(1)$$
$$= \frac{|\mathbf{G}^F|_{p'}}{|C_{\mathbf{G}^*}(s)^{F^*}|_{p'}} \psi_s(\chi)(1),$$

the last equality by (1) applied in $C_{\mathbf{G}^*}(s)^{F^*}$. ■

We will be able here to give explicitly the bijection of 13.23 when $C_{\mathbf{G}^*}(s)$ is a Levi subgroup of a parabolic subgroup of $\mathbf{G}^*$; this is in a sense the "general case" as "most" semi-simple elements have this property: we will show (see just after 14.11) that, modulo the centre, there is only a finite number of semi-simple elements whose centralizer is not in a proper Levi subgroup of $\mathbf{G}$.

To do that, we must first explain the correspondence between rational Levi subgroups of $\mathbf{G}$ and rational Levi subgroups of $\mathbf{G}^*$. We first remark that, as the parametrization given by 4.3 is the same in $\mathbf{G}$ and $\mathbf{G}^*$, rational classes of rational Levi subgroups of $\mathbf{G}$ correspond one-to-one to those of $\mathbf{G}^*$. If the class of $\mathbf{L}$ corresponds to the F-class under $W(\mathbf{T})$ of $W_I w$, *i.e.*, to the $W(\mathbf{T})$-class of $W_I wF$, we make $\mathbf{L}$ correspond to the Levi subgroup $\mathbf{L}^*$ of $\mathbf{G}^*$ whose class is parametrized by $F^* w^* W_I^*$. We may find in $\mathbf{L}$ a torus $\mathbf{T}_w$ of type w, *i.e.*, $(\mathbf{T}_w, F)$ is geometrically conjugate to $(\mathbf{T}, wF)$; similarly we may find in $\mathbf{L}^*$ a torus $\mathbf{T}_w^*$ such that $(\mathbf{T}_w^*, F^*)$ is geometrically conjugate to $(\mathbf{T}^*, F^* w^*)$. With these choices $(\mathbf{L}, \mathbf{T}_w, F)$ is dual to $(\mathbf{L}^*, \mathbf{T}_w^*, F^*)$, where the isomorphism $Y(\mathbf{T}_w) \xrightarrow{\sim} X(\mathbf{T}_w^*)$ is "transported" from the isomorphism $Y(\mathbf{T}) \xrightarrow{\sim} X(\mathbf{T}^*)$ by the chosen geometric conjugations. Given the duality between $(\mathbf{G}, \mathbf{T})$ and $(\mathbf{G}^*, \mathbf{T}^*)$, the duality thus defined between $\mathbf{L}$ and $\mathbf{L}^*$ is defined up to the automorphisms of $\mathbf{L}$ induced by $\mathbf{G}$, *i.e.*, by $N_{\mathbf{G}}(\mathbf{L})$. It is easy to check that this duality is compatible with 13.12 and 13.13, *i.e.*, if $(\mathbf{T}, \theta)$ where $\mathbf{T} \subset \mathbf{L}$ corresponds to $(\mathbf{T}^*, s)$ where $\mathbf{T}^* \subset \mathbf{L}^*$ for this duality, then it also corresponds to $(\mathbf{T}^*, s)$ for the duality in $\mathbf{G}$.

We may then state the following result, from which we will deduce 13.23 when $C_{\mathbf{G}^*}(s)$ is a Levi subgroup.

13.25 THEOREM. *Let s be a semi-simple element of $\mathbf{G}^{*F}$ and let $\mathbf{L}^*$ be a rational Levi subgroup of $\mathbf{G}^*$ which contains $C_{\mathbf{G}^*}(s)$. Let $\mathbf{L}$ be a Levi* *subgroup of* $\mathbf{G}$ *whose* $\mathbf{G}^F$*-class corresponds to* $\mathbf{L}^*$ *as explained above, and let* $\mathbf{LU}$ *be the Levi decomposition of some parabolic subgroup of* $\mathbf{G}$ *containing* $\mathbf{L}$*; then:*

(i) *For any* $\pi \in \mathcal{E}(\mathbf{L}^F, (s))$ *the space* $H^i(\mathcal{L}^{-1}(\mathbf{U})) \otimes_{\overline{\mathbb{Q}}_\ell[\mathbf{L}^F]} \pi$ *is zero for* $i \neq d$ *where* $d = \dim \mathbf{U} + \dim(\mathbf{U} \cap {}^F\mathbf{U})$ *and affords an irreducible representation of* $\mathbf{G}^F$ *for* $i = d$.

(ii) *The functor* $\varepsilon_{\mathbf{G}}\varepsilon_{\mathbf{L}} R_{\mathbf{L}}^{\mathbf{G}}$ *induces a bijection from* $\mathcal{E}(\mathbf{L}^F, (s))$ *to* $\mathcal{E}(\mathbf{G}^F, (s))$.

PROOF: We begin with a lemma which translates for the group $\mathbf{G}$ the condition $C_{\mathbf{G}^*}(s) \subset \mathbf{L}^*$.

13.26 LEMMA.

(i) *Let* $\mathbf{T}$ *and* $\mathbf{T}'$ *be two rational maximal tori of* $\mathbf{L}$ *and let* $\theta \in \operatorname{Irr}(\mathbf{T}^F)$ *and* $\theta' \in \operatorname{Irr}(\mathbf{T}'^F)$ *be such that* $(\mathbf{T}, \theta)$ *and* $(\mathbf{T}', \theta')$ *are geometrically conjugate in* $\mathbf{L}$ *and their geometric class corresponds to the class of* s *in* $\mathbf{L}^*$ *by the duality between* $\mathbf{L}$ *and* $\mathbf{L}^*$*; then any* $g \in \mathbf{G}$ *which geometrically conjugates* $(\mathbf{T}, \theta)$ *to* $(\mathbf{T}', \theta')$ *is in* $\mathbf{L}$.

(ii) *Let* $\mathbf{T}$*, a rational maximal torus of* $\mathbf{G}$ *and* $\theta \in \operatorname{Irr}(\mathbf{T}^F)$ *be such that the geometric conjugacy class of* $(\mathbf{T}, \theta)$ *corresponds to the* $\mathbf{G}^*$*-class of* s*. Then* $\mathbf{T}$ *is* $\mathbf{G}^F$*-conjugate to some torus of* $\mathbf{L}$.

PROOF: To prove (i) it is clearly enough to consider the case where $\mathbf{T} = \mathbf{T}'$ and $\theta = \theta'$. Up to geometric conjugacy, we may even "transport" the situation to $\mathbf{T}_w$. Thus we need only consider elements $g \in N_{\mathbf{G}}(\mathbf{T}_w)$ which fix some character $\theta_0 \in \operatorname{Irr}(\mathbf{T}_w^{vF})$ for some $v \in W(\mathbf{T}_w)$ where θ_0 is such that the semi-simple element $s_0 \in \mathbf{T}_w^*$ corresponding to θ_0 is geometrically conjugate to s in $\mathbf{L}^*$. In particular we have $C_{\mathbf{G}^*}(s_0) \subset \mathbf{L}^*$, so $C_{W(\mathbf{T}_w^*)}(s) \subset W_{\mathbf{L}^*}(\mathbf{T}_w)$. By the anti-isomorphism between $W(\mathbf{T}_w)$ and $W(\mathbf{T}_w^*)$ the element g corresponds to an element of $C_{W(\mathbf{T}_w^*)}(s)$, whence the result.

We now prove (ii). Let $\mathbf{T}_0 \subset \mathbf{L}$ and $\theta_0 \in \operatorname{Irr}(\mathbf{T}_0^F)$ be such that the geometric class in $\mathbf{L}$ of the pair $(\mathbf{T}_0, \theta_0)$ corresponds to the class in $\mathbf{L}^*$ of s. Then $(\mathbf{T}_0, \theta_0)$ also corresponds to (s) for the duality between $\mathbf{G}$ and $\mathbf{G}^*$, so $(\mathbf{T}, \theta)$ and $(\mathbf{T}_0, \theta_0)$ are geometrically conjugate in $\mathbf{G}$. Let $g \in \mathbf{G}$ be an element which geometrically conjugates $(\mathbf{T}, \theta)$ to $(\mathbf{T}_0, \theta_0)$. Then ${}^F g$ also geometrically conjugates $(\mathbf{T}, \theta)$ to $(\mathbf{T}_0, \theta_0)$, so $g.{}^F g^{-1}$ geometrically conjugates $(\mathbf{T}_0, \theta_0)$ to itself. Thus by (i) we have $g.{}^F g^{-1} \in \mathbf{L}$. Using the Lang-Steinberg theorem to write $g.{}^F g^{-1} = l^{-1}.{}^F l$ with $l \in \mathbf{L}$, we get an element lg which is rational and conjugates $\mathbf{T}$ to some torus of $\mathbf{L}$. ■

We now place ourselves in the setting of 11.6, where we take $\mathbf{L}' = \mathbf{L}$ and $\mathbf{Q} = \mathbf{P} = \mathbf{LU}$. The key step for 13.25 is

13.27 PROPOSITION. *Under the same assumptions as 13.25, let* π *and* π' *be two irreducible representations in* $\mathcal{E}(\mathbf{L}^F,(s))$*, then*

$$\pi^\vee \otimes_{\overline{\mathbb{Q}}_\ell[\mathbf{L}^F]} H^i_c(\mathbf{Z},\overline{\mathbb{Q}}_\ell) \otimes_{\overline{\mathbb{Q}}_\ell[\mathbf{L}^F]} \pi'$$

is 0 except for $i = 2d$*, where* $d = \dim \mathbf{U} + \dim(\mathbf{U} \cap {}^F\mathbf{U})$ *and* $\pi^\vee$ *is as in the proof of 13.3. In addition we have*

$$\dim(\pi^\vee \otimes_{\overline{\mathbb{Q}}_\ell[\mathbf{L}^F]} H^{2d}_c(\mathbf{Z},\overline{\mathbb{Q}}_\ell) \otimes_{\overline{\mathbb{Q}}_\ell[\mathbf{L}^F]} \pi') = \langle \pi, \pi' \rangle_{\mathbf{L}^F}.$$

PROOF: We may choose two rational maximal tori $\mathbf{T}$ and $\mathbf{T}'$ in $\mathbf{L}$, characters $\theta \in \mathrm{Irr}(\mathbf{T}^F)$ and $\theta' \in \mathrm{Irr}(\mathbf{T}'^F)$ and $\mathbf{TV}$ (resp. $\mathbf{T}'\mathbf{V}'$) (the Levi decomposition of) a Borel subgroup of $\mathbf{L}$ containing $\mathbf{T}$ (resp. $\mathbf{T}'$) such that there exists j (resp. k) for which $\langle \pi, H^j_c(\mathcal{L}^{-1}(\mathbf{V})) \otimes_{\overline{\mathbb{Q}}_\ell[\mathbf{T}^F]} \theta \rangle_{\mathbf{L}^F} \neq 0$ (resp. $\langle \pi', H^k_c(\mathcal{L}^{-1}(\mathbf{V}')) \otimes_{\overline{\mathbb{Q}}_\ell[\mathbf{T}'^F]} \theta' \rangle_{\mathbf{L}^F} \neq 0$); then $\pi^\vee \otimes_{\overline{\mathbb{Q}}_\ell[\mathbf{L}^F]} H^i_c(\mathbf{Z}_w,\overline{\mathbb{Q}}_\ell) \otimes_{\overline{\mathbb{Q}}_\ell[\mathbf{L}^F]} \pi'$ is a subspace of

$$\theta^\vee \otimes_{\overline{\mathbb{Q}}_\ell[\mathbf{T}^F]} H^j_c(\mathcal{L}^{-1}(\mathbf{V})^\vee) \otimes_{\overline{\mathbb{Q}}_\ell[\mathbf{L}^F]} H^i_c(\mathbf{Z}_w,\overline{\mathbb{Q}}_\ell) \otimes_{\overline{\mathbb{Q}}_\ell[\mathbf{L}^F]} H^k_c(\mathcal{L}^{-1}(\mathbf{V}')) \otimes_{\overline{\mathbb{Q}}_\ell[\mathbf{T}'^F]} \theta'.$$

By similar arguments to those used in the proof of 11.5, the $\mathbf{T}^F$-module-$\mathbf{T}'^F$ given by

$$H^j_c(\mathcal{L}^{-1}(\mathbf{V})^\vee) \otimes_{\overline{\mathbb{Q}}_\ell[\mathbf{L}^F]} H^i_c(\mathbf{Z}_w,\overline{\mathbb{Q}}_\ell) \otimes_{\overline{\mathbb{Q}}_\ell[\mathbf{L}^F]} H^k_c(\mathcal{L}^{-1}(\mathbf{V}'))$$

is isomorphic to a submodule of $H^{i+j+k}_c(\bigcup_{w_1} \mathbf{Z}^1_{w_1},\overline{\mathbb{Q}}_\ell)$, where $\mathbf{Z}^1_{w_1}$ is the variety analogous to $\mathbf{Z}_w$, relative to $\mathbf{T}$ and $\mathbf{T}'$, and where w_1 runs over elements of $\mathbf{T}\backslash\mathcal{S}(\mathbf{T},\mathbf{T}')/\mathbf{T}'$ having the same image in $\mathbf{L}\backslash\mathcal{S}(\mathbf{L},\mathbf{L})/\mathbf{L}$ as w. By an argument similar to the proofs of 13.4 and 13.5, the character $\theta \otimes \theta'^\vee$ does not occur in this module if none of the w_1 geometrically conjugates θ to θ'. By 13.26 (i), this implies that w is in $\mathbf{L}$. Thus $\pi^\vee \otimes_{\overline{\mathbb{Q}}_\ell[\mathbf{L}^F]} H^i_c(\mathbf{Z}_w,\overline{\mathbb{Q}}_\ell) \otimes_{\overline{\mathbb{Q}}_\ell[\mathbf{L}^F]} \pi'$ is zero if w is not in $\mathbf{L}$. If w is in $\mathbf{L}$, *i.e.*, $w = 1$ in $\mathbf{L}\backslash\mathcal{S}(\mathbf{L},\mathbf{L})/\mathbf{L}$, the variety $\mathbf{Z}_1$ (see definition of $\mathbf{Z}_w$ after 11.7) may be simplified. We have

$$\mathbf{Z}_1 = \{(u,u',g) \in \mathbf{U} \times \mathbf{U} \times {}^{F^{-1}}\mathbf{P} \mid u\,{}^F g = g u'\},$$

so $(u,u',g) \in \mathbf{Z}_1$ implies $u\,{}^F g u'^{-1} = g \in \mathbf{P} \cap {}^{F^{-1}}\mathbf{P}$. Using the decomposition $\mathbf{P} \cap {}^{F^{-1}}\mathbf{P} = \mathbf{L}.(\mathbf{U} \cap {}^{F^{-1}}\mathbf{U})$, and taking projections to $\mathbf{L}$, we see that $g \in \mathbf{L}^F.(\mathbf{U} \cap {}^{F^{-1}}\mathbf{U})$; so the variety projects to $\mathbf{L}^F$ with all fibres isomorphic to $\mathbf{U} \times (\mathbf{U} \cap {}^F\mathbf{U})$. Thus, up to a shift of $2d$, the cohomology is that of the discrete variety $\mathbf{L}^F$, thus

$$\pi^\vee \otimes_{\overline{\mathbb{Q}}_\ell[\mathbf{L}^F]} H^i_c(\mathbf{Z}_w,\overline{\mathbb{Q}}_\ell) \otimes_{\overline{\mathbb{Q}}_\ell[\mathbf{L}^F]} \pi' = \begin{cases} \pi^\vee \otimes_{\overline{\mathbb{Q}}_\ell[\mathbf{L}^F]} \overline{\mathbb{Q}}_\ell[\mathbf{L}^F] \otimes_{\overline{\mathbb{Q}}_\ell[\mathbf{L}^F]} \pi' & \text{if } w \in \mathbf{L} \text{ and } i = 2d, \\ 0 & \text{otherwise,} \end{cases}$$

and the dimension of $\pi^\vee \otimes_{\overline{\mathbb{Q}}_\ell[\mathbf{L}^F]} \overline{\mathbb{Q}}_\ell[\mathbf{L}^F] \otimes_{\overline{\mathbb{Q}}_\ell[\mathbf{L}^F]} \pi'$ is $\langle \pi, \pi' \rangle_{\mathbf{L}^F}$. Using the arguments of 13.5 we get the required result for $\pi^\vee \otimes_{\overline{\mathbb{Q}}_\ell[\mathbf{L}^F]} H^i_c(\mathbf{Z}, \overline{\mathbb{Q}}_\ell) \otimes_{\overline{\mathbb{Q}}_\ell[\mathbf{L}^F]} \pi'$. ■

As the scalar product $\langle R^{\mathbf{G}}_{\mathbf{L}}\pi, R^{\mathbf{G}}_{\mathbf{L}}\pi' \rangle_{\mathbf{G}^F}$ is equal to the dimension of

$$\pi^\vee \otimes_{\overline{\mathbb{Q}}_\ell[\mathbf{L}^F]} H^*(\mathcal{L}^{-1}(\mathbf{U})^\vee) \otimes_{\overline{\mathbb{Q}}_\ell[\mathbf{G}^F]} H^*(\mathcal{L}^{-1}(\mathbf{U})) \otimes_{\overline{\mathbb{Q}}_\ell[\mathbf{L}^F]} \pi',$$

the above proposition thus shows that

$$\langle R^{\mathbf{G}}_{\mathbf{L}}\pi, R^{\mathbf{G}}_{\mathbf{L}}\pi' \rangle_{\mathbf{G}^F} = \langle \pi, \pi' \rangle_{\mathbf{L}^F}.$$

13.28 REMARK. We note that in the above situation, even if we do not assume that $\mathbf{P} = \mathbf{P}'$, the argument of the preceding proof (that $w \in \mathbf{L}$) and lemma 11.12 show that the Mackey formula holds for π and π', *i.e.*,

$$\langle R^{\mathbf{G}}_{\mathbf{L}\subset\mathbf{P}}\pi, R^{\mathbf{G}}_{\mathbf{L}\subset\mathbf{P}'}\pi' \rangle_{\mathbf{G}} = \langle \pi, \pi' \rangle_{\mathbf{L}^F}.$$

As seen in 6.1.1, this implies that in this case $R^{\mathbf{G}}_{\mathbf{L}\subset\mathbf{P}}\pi$ is independent of $\mathbf{P}$.

We now prove theorem 13.25. Suppose that the $\mathbf{G}^F$-module given by

$$H^j(\mathcal{L}^{-1}(\mathbf{U})) \otimes_{\overline{\mathbb{Q}}_\ell[\mathbf{L}^F]} \pi$$

is not 0. Let χ be one of its irreducible components. Then $\chi^\vee$ occurs in $\pi^\vee \otimes_{\overline{\mathbb{Q}}_\ell[\mathbf{L}^F]} H^j_c(\mathcal{L}^{-1}(\mathbf{U})^\vee)$ and thus

$$\pi^\vee \otimes_{\overline{\mathbb{Q}}_\ell[\mathbf{L}^F]} H^j_c(\mathcal{L}^{-1}(\mathbf{U})^\vee) \otimes_{\overline{\mathbb{Q}}_\ell[\mathbf{G}^F]} H^j(\mathcal{L}^{-1}(\mathbf{U})) \otimes_{\overline{\mathbb{Q}}_\ell[\mathbf{L}^F]} \pi \neq 0.$$

But this is a subspace of $\pi^\vee \otimes_{\overline{\mathbb{Q}}_\ell[\mathbf{L}^F]} H^{2j}_c(\mathbf{Z}, \overline{\mathbb{Q}}_\ell) \otimes_{\overline{\mathbb{Q}}_\ell[\mathbf{L}^F]} \pi'$, so this last space is not 0, which proves by 13.27 that $j = d$. Since in that case the last space is of dimension at most 1, we see that χ must be in addition the only irreducible component of $H^j(\mathcal{L}^{-1}(\mathbf{U})) \otimes_{\overline{\mathbb{Q}}_\ell[\mathbf{L}^F]} \pi$. Whence (i) of the theorem.

Let us show (ii). (i) implies that $R^{\mathbf{G}}_{\mathbf{L}}$ induces an isometry from $\mathcal{E}(\mathbf{L}^F, (s))$ to $\mathcal{E}(\mathbf{G}^F, (s))$. Let us show that this isometry is surjective. Let $\chi \in \mathcal{E}(\mathbf{G}^F, (s))$; by definition χ is a component of $R^{\mathbf{G}}_{\mathbf{T}}(\theta')$ for some torus $\mathbf{T}'$ and some character θ' which is in the class defined by s. By 13.26 (ii), the torus $\mathbf{T}'$ has a rational conjugate in $\mathbf{L}$. We may thus assume that $\mathbf{T}'$ is in $\mathbf{L}$. In that case $R^{\mathbf{G}}_{\mathbf{T}'}(\theta) = R^{\mathbf{G}}_{\mathbf{L}} \circ R^{\mathbf{L}}_{\mathbf{T}'}(\theta)$, so χ is indeed the $R^{\mathbf{G}}_{\mathbf{L}}$ of a character of $\mathcal{E}(\mathbf{L}^F, (s))$.

It remains to see that $\varepsilon_{\mathbf{G}}\varepsilon_{\mathbf{L}} R^{\mathbf{G}}_{\mathbf{L}}$ maps a true character to a true character. This follows immediately from 12.17. ■

13.29 EXERCISE.

1. With the notation of the previous theorem, show that if w is the type of a quasi-split torus of $\mathbf{L}$ with respect to a quasi-split torus of $\mathbf{G}$, then $l(w) = d$.
2. Use this result and (i) of the theorem to show directly that $\varepsilon_{\mathbf{G}}\varepsilon_{\mathbf{L}}R_{\mathbf{L}}^{\mathbf{G}}$ maps a true character to a true character (hint: we have $\varepsilon_{\mathbf{G}}\varepsilon_{\mathbf{L}} = (-1)^{l(w)}$).

The special case of theorem 13.25 where $\mathbf{L}^* = C_{\mathbf{G}^*}(s)$ shows that $R_{\mathbf{L}}^{\mathbf{G}}$ is then a bijection from $\mathcal{E}(C_{\mathbf{G}^*}(s)^{*F}, (s))$ to $\mathcal{E}(\mathbf{G}^F, (s))$. However

13.30 PROPOSITION. *Let s be a central element of $\mathbf{G}^{*F^*}$. Then:*

(i) There exists a linear character $\hat{s} \in \mathrm{Irr}(\mathbf{G}^F)$ such that, for any rational maximal torus $\mathbf{T}$ of $\mathbf{G}$, the pair $(\mathbf{T}, \hat{s}|_{\mathbf{T}^F})$ is in the geometric conjugacy class defined by s.

(ii) Taking the tensor product with $\hat{s}$ defines a bijection

$$\mathcal{E}(\mathbf{G}^F, 1) \xrightarrow{\sim} \mathcal{E}(\mathbf{G}^F, (s)).$$

PROOF: As $s \in Z(\mathbf{G}^{*F^*})$, its geometric class defines a character of the rational points of any rational maximal torus of $\mathbf{G}$ since, if $(\mathbf{G}, \mathbf{T})$ is dual to $(\mathbf{G}^*, \mathbf{T}^*)$, the element s is in $\mathbf{T}^{*wF^*}$ for any w. Let θ be the character of $\mathbf{T}^F$ which corresponds to s. As any root of $\mathbf{G}^*$ relative to $\mathbf{T}$ vanishes on s, we get that θ is trivial on the rational points of the subtorus generated by the images of the coroots. If $\mathbf{T}'$ is a maximal torus in a semi-simple group, by 0.37 (iii) we have $<\Phi^\vee> \otimes \mathbb{Q} = Y(\mathbf{T}') \otimes \mathbb{Q}$, so, using 0.20 $\mathbf{T}'$ is generated by the images of the coroots; so in general the subtorus generated by the images of the coroots is $\mathbf{T} \cap \mathbf{G}'$ where $\mathbf{G}'$ is the derived group of $\mathbf{G}$. Thus θ factors through $\mathbf{T}^F/(\mathbf{T}^F \cap \mathbf{G}')$. As the restriction to $\mathbf{T}$ of the quotient morphism $\mathbf{G} \to \mathbf{G}/\mathbf{G}'$ is surjective (as $\mathbf{G} = \mathbf{G}'.R(\mathbf{G})$, and $R(\mathbf{G})$ is in all tori), we have $\mathbf{T}/(\mathbf{T} \cap \mathbf{G}') \simeq \mathbf{G}/\mathbf{G}'$; as both $\mathbf{T} \cap \mathbf{G}'$ and $\mathbf{G}'$ are connected, by the Lang-Steinberg theorem the isomorphism carries over to the rational points: we get $\mathbf{T}^F/(\mathbf{T}^F \cap \mathbf{G}') \simeq \mathbf{G}^F/(\mathbf{G}')^F$. Thus s defines some character $\hat{s}$ of the abelian group $\mathbf{G}^F/(\mathbf{G}')^F$. The character thus defined is independent of the torus used for its construction, since the characters obtained in various tori are geometrically conjugate, and geometric conjugacy is the identity on $\mathbf{G}^F/(\mathbf{G}')^F$ (since it is trivial on $(\mathbf{G}/\mathbf{G}')^F \simeq \mathbf{G}^F/\mathbf{G}'^F$). Let us now prove (ii). Taking the tensor product with $\hat{s}$ permutes the irreducible characters, and by 12.2 we have $R_{\mathbf{T}}^{\mathbf{G}}(\theta)\hat{s} = R_{\mathbf{T}}^{\mathbf{G}}(\theta\hat{s})$ for any $\mathbf{T}$. As the isomorphism $\mathrm{Irr}(\mathbf{T}^{wF}) \simeq \mathbf{T}^{*F^*w^*}$ is compatible with multiplication, if the geometric class of $(\mathbf{T}, \theta)$ corresponds to that of $s_1 \in \mathbf{G}^{*F^*}$, then the class of $(\mathbf{T}, \theta\hat{s})$ corresponds to that of $s_1 s$, whence (ii). ■

13.31 REMARK. Actually it can be shown that $\mathbf{G}^F/(\mathbf{G}')^F$ is the semi-simple part of the abelian quotient of $\mathbf{G}^F$ (for quasi-simple groups, the abelian quotient of $\mathbf{G}^F$ is semi-simple except for $A_1(\mathbb{F}_2)$, $A_1(\mathbb{F}_3)$, $B_2(\mathbb{F}_2)$, ${}^2A_2(\mathbb{F}_2)$ and $G_2(\mathbb{F}_2)$; see [J. TITS, Algebraic simple groups and abstract groups, *Ann. Math.* **80** (1964), 313–329]).

The above proposition together with 13.25 (ii) gives the special case of 13.23 where $C_{\mathbf{G}^*}(s)$ is a Levi subgroup, if we show that the unipotent characters of a group and of its dual are in one-to-one correspondence; this results from 13.20 when the root systems of $\mathbf{G}$ and $\mathbf{G}^*$ are of the same type, *i.e.*, when $\mathbf{G}$ has no quasi-simple components of type B_n or C_n; it was proved in this last case by Lusztig in [L3].

References

The statements in this chapter about geometric conjugacy of $(\mathbf{T}, \theta)$ and its interpretation in the dual group are in Deligne and Lusztig [DL1]. Langlands had introduced before a dual group which has the same root system as $\mathbf{G}^*$ but is over the complex field. Most of 13.25 is already in Lusztig [L1]. In subsequent works ([Coxeter orbits and eigenspaces of Frobenius, *Inventiones Math.*, **28** (1976), 101–159], [L3], [L4]) Lusztig first gave the definition of series, and then parametrized, initially unipotent characters, then all characters, and subsequently gave the decomposition of Deligne-Lusztig characters into irreducible components. Lusztig's parametrization of unipotent characters is described in [Ca].

14. REGULAR ELEMENTS; GELFAND-GRAEV REPRESENTATIONS; REGULAR AND SEMI-SIMPLE CHARACTERS

The aim of this chapter is to expound the main results about Gelfand-Graev representations. As the properties of these representations involve the notion of regular elements, we begin with their definition and some of their properties, in particular of regular unipotent elements.

14.1 DEFINITION. *An element x of an algebraic group* $\mathbf{G}$ *is said to be* **regular** *if the dimension of its centralizer is minimal.*

If $\mathbf{G}$ is reductive, this dimension is actually equal to the rank of $\mathbf{G}$ by the following result.

14.2 PROPOSITION. *Let* $\mathbf{G}$ *be a connected linear algebraic group.*

(i) For any element $x \in \mathbf{G}$ we have $\dim C_{\mathbf{G}}(x) \geq \mathrm{rk}(\mathbf{G})$.

(ii) Assume $\mathbf{G}$ *reductive; then in any torus* $\mathbf{T}$ *of* $\mathbf{G}$ *there exists an element t such that $C_{\mathbf{G}}(t) = \mathbf{T}$.*

PROOF: Let $\mathbf{B}$ be a Borel subgroup of $\mathbf{G}$ containing x and let $\mathbf{U}$ be its unipotent radical. The conjugacy class of x under $\mathbf{B}$ is contained in $x\mathbf{U}$, so its dimension is at most $\dim(\mathbf{U}) = \dim(\mathbf{B}) - \mathrm{rk}(\mathbf{G})$. As this dimension is equal to $\dim(\mathbf{B}) - \dim C_{\mathbf{B}}(x)$, we get $\dim C_{\mathbf{B}}(x) \geq \mathrm{rk}(\mathbf{G})$, whence $\dim C_{\mathbf{G}}(x) \geq \mathrm{rk}(\mathbf{G})$. We now prove (ii). If we apply 0.7 to the action of $\mathbf{T}$ on $\mathbf{G}$ by conjugation, we get an element $t \in \mathbf{T}$ such that $C_{\mathbf{G}}(t) = C_{\mathbf{G}}(\mathbf{T})$. But by 0.32 this means that $C_{\mathbf{G}}(t) = \mathbf{T}$, whence the result. ■

14.3 EXAMPLES. Take $\mathbf{G} = \mathbf{GL}_n$; its rank is n (see chapter 15). Any diagonal matrix with diagonal entries all distinct is regular, as its centralizer is the torus which consists of all diagonal matrices. The unipotent matrix

$$\begin{pmatrix} 1 & 1 & & \\ & 1 & \ddots & \\ & & \ddots & 1 \\ & & & 1 \end{pmatrix}$$

is regular, as its centralizer is the group of all upper triangular matrices with constant diagonals.

From now on $\mathbf{G}$ will denote a reductive group. The following result gives a characterization of regular elements.

14.4 THEOREM. *Let $x = su$ be the Jordan decomposition of an element of* $\mathbf{G}$. *The element x is regular if and only if u is regular in $C^\circ_{\mathbf{G}}(s)$.*

PROOF: We have $C_{\mathbf{G}}(su) = C_{\mathbf{G}}(s) \cap C_{\mathbf{G}}(u)$, so $\dim C_{\mathbf{G}}(su) = \dim C_{C^\circ_{\mathbf{G}}(s)}(u)$. But su is regular if and only if this dimension is equal to $\mathrm{rk}(\mathbf{G}) = \mathrm{rk}(C^\circ_{\mathbf{G}}(s))$. By 14.2 applied in $C^\circ_{\mathbf{G}}(s)$ this is equivalent to u being regular in $C^\circ_{\mathbf{G}}(s)$. ■

Note that by 14.2 we know that regular semi-simple elements exist. The existence of regular unipotent elements is much more difficult to prove, and will be done later. Once this existence is known, theorem 14.4 shows that there are regular elements with arbitrary semi-simple parts.

14.5 EXERCISE. Show that in $\mathbf{GL}_n$ or $\mathbf{SL}_n$ a matrix is regular if and only if its characteristic polynomial is equal to its minimal polynomial.

We first study semi-simple regular elements.

14.6 PROPOSITION.

(i) *Let $\mathbf{T}$ be a maximal torus of $\mathbf{G}$; an element s of $\mathbf{T}$ is regular if and only if no root of $\mathbf{G}$ relative to $\mathbf{T}$ is trivial on s.*

(ii) *A semi-simple element is regular if and only if it is contained in only one maximal torus.*

(iii) *A semi-simple element s is regular if and only if $C^\circ_{\mathbf{G}}(s)$ is a torus.*

PROOF: Assertions (i) and (iii) are straightforward from proposition 2.3. Assertion (ii) is a direct consequence of assertion (iii). ■

14.7 COROLLARY. *Regular semi-simple elements are dense in any Borel subgroup.*

PROOF: Let $\mathbf{B}$ be a Borel subgroup and $\mathbf{T}$ be a maximal torus of $\mathbf{B} = \mathbf{TU}$; if $t \in \mathbf{T}$ is regular and u is any element of $\mathbf{U}$, then the semi-simple part of tu is conjugate to t (see 7.1), so is also regular. But an element whose semi-simple part is regular has to be semi-simple as, by 14.6 (iii), it is then in a torus (see 2.5), so tu is regular semi-simple. By 14.6 (i) the regular elements of $\mathbf{T}$ form an open (and so dense) subset $\mathbf{T}_{\mathrm{reg}}$ of $\mathbf{T}$. So the set $\mathbf{T}_{\mathrm{reg}} \times \mathbf{U}$, which consists of regular semi-simple elements, is dense in $\mathbf{B}$. ■

14.8 REMARK. The above proof shows that if $t \in \mathbf{T}$ is regular then the map $u \mapsto u^{-1}\,{}^t u$ is surjective from $\mathbf{U}$ onto itself.

14.9 COROLLARY. *Regular semi-simple elements are dense in* $\mathbf{G}$.

PROOF: This is clear from 14.7 as $\mathbf{G}$ is the union of all Borel subgroups. ■

We now consider regular unipotent elements. The proof of their existence (due to Steinberg) is not easy. We shall prove it only for groups over $\overline{\mathbb{F}}_q$

by deducing it from the fact that there is only a finite number of unipotent classes in $\mathbf{G}$. This last result is true in any characteristic (the most difficult case being finite characteristic). We follow here a proof given by Lusztig, [L1], for the case of groups over $\overline{\mathbb{F}}_q$ because it is a good illustration of the theory of representations of finite groups of Lie type. In that proof the group $\mathbf{G}$ is assumed to have a connected centre. As the quotient morphism $\mathbf{G} \to \mathbf{G}/Z(\mathbf{G})$ induces a bijection on unipotent elements and also on unipotent classes this assumption does not restrict the generality of the result. The idea of the proof is to use the representation theory of $\mathbf{G}^F$ to separate the unipotent classes of $\mathbf{G}^F$ by a finite set of class functions whose cardinality is bounded independently of q, which gives the result as $\mathbf{G} = \bigcup_{n \in \mathbb{N}^*} \mathbf{G}^{F^n}$.

First we need some notation. If $\mathbf{G}$ is a connected reductive group over $\overline{\mathbb{F}}_q$, defined over $\mathbb{F}_q$, and $\mathbf{G}^*$ is dual to $\mathbf{G}$ (we shall denote by $\mathbf{T}$ and $\mathbf{T}^*$ a pair of dual rational maximal tori) we define $\mathcal{E}(\mathbf{G}^F)$ to be the linear span of the Lusztig series $\mathcal{E}(\mathbf{G}^F, (s))$ when s runs over the set $\mathcal{S}(\mathbf{G}^*)$ of semi-simple elements of $\mathbf{G}^{*F^*}$ such that there exists a rational Levi subgroup $\mathbf{L}^*$ of $\mathbf{G}^*$ whose derived group $\mathbf{L}^{*\prime}$ contains s and such that $C_{\mathbf{L}^*}(s)$ is not contained in some proper rational Levi subgroup of $\mathbf{L}^*$. We denote by $\mathcal{E}(\mathbf{G}^F)_u$ the space of restrictions of $\mathcal{E}(\mathbf{G}^F)$ to unipotent elements. The main lemma is the following.

14.10 LEMMA. *Let $\mathbf{G}$ be a connected reductive group over $\overline{\mathbb{F}}_q$ with connected centre and let F be a Frobenius endomorphism on $\mathbf{G}$ defining an $\mathbb{F}_q$-structure; the restriction of any class function f on $\mathbf{G}^F$ to unipotent elements is in $\mathcal{E}(\mathbf{G}^F)_u$.*

PROOF: The proof is by induction on the dimension of $\mathbf{G}$. If $Z(\mathbf{G}) \neq \{1\}$, the map $\mathbf{G} \to \mathbf{G}/Z(\mathbf{G})$ induces a bijection on rational unipotent elements, so the restriction of f to unipotent elements is the restriction of a class function $\overline{f}$ on $(\mathbf{G}/Z(\mathbf{G}))^F$ which group is equal to $\mathbf{G}^F/Z(\mathbf{G})^F$ since $Z(\mathbf{G})$ is connected. By the induction hypothesis we have $\overline{f} \in \mathcal{E}(\mathbf{G}^F/Z(\mathbf{G})^F)_u$. But the dual group of $\mathbf{G}/Z(\mathbf{G})$ is isomorphic to a subgroup of $\mathbf{G}^*$: the subgroup generated by the root subgroups and a subtorus of $\mathbf{T}^*$ dual to $\mathbf{T}/Z(\mathbf{G})$. As the dual of a semi-simple group the group $(\mathbf{G}/Z(\mathbf{G}))^*$ is semi-simple, so is contained in $(\mathbf{G}^*)'$. Whence $f \in \mathcal{E}(\mathbf{G}^F)_u$ in this case. We now assume that $Z(\mathbf{G}) = 1$, so $\mathbf{G}$ and $\mathbf{G}^*$ are semi-simple. Let ρ be an irreducible character of $\mathbf{G}^F$; it is in some Lusztig series $\mathcal{E}(\mathbf{G}^F, (s))$. Assume first that the centralizer of s is contained in some proper rational Levi subgroup $\mathbf{L}^*$ of $\mathbf{G}^*$. Then, by 13.25 (ii), we have $\rho = R_{\mathbf{L}}^{\mathbf{G}}(\pi)$, where π is, up to sign, an element of $\mathcal{E}(\mathbf{L}^F, (s))$. By the induction hypothesis applied to $\mathbf{L}$, the restriction of π to the unipotent elements of $\mathbf{L}$ is in $\mathcal{E}(\mathbf{L}^F)_u$. As $R_{\mathbf{L}}^{\mathbf{G}}$ commutes with the restriction to the unipotent elements (see 12.6) and maps $\mathcal{E}(\mathbf{L}^F)$ to $\mathcal{E}(\mathbf{G}^F)$, the restriction of $R_{\mathbf{L}}^{\mathbf{G}}(\pi)$ to the unipotent elements is in $\mathcal{E}(\mathbf{G}^F)_u$, whence the result in this case. Assume now that the centralizer of s is not contained in a proper rational Levi subgroup. We have $s \in (\mathbf{G}^*)' = \mathbf{G}^*$, so $\mathcal{E}(\mathbf{G}^F, (s)) \subset \mathcal{E}(\mathbf{G}^F)$ and we get the result in this last case. ■

We now have to show that the dimension of $\mathcal{E}(\mathbf{G}^F)$ is bounded independently of q. By 13.14 (iii), in the dual of a group with connected centre the semi-simple elements have connected centralizers. So the condition that the centralizer of s is not contained in a proper Levi subgroup of some Levi subgroup $\mathbf{L}^*$ can be seen on the root system of $C_{\mathbf{G}^*}(s)$ (with respect to some torus containing s). Explicitly, an element $s \in \mathbf{G}^{*F^*}$ is in $\mathcal{S}(\mathbf{G}^*)$ if and only if there exists a Levi subgroup $\mathbf{L}^*$ such that $s \in \mathbf{L}^{*\prime}$ and that the root system of $C_{\mathbf{L}^*}(s)$ is not contained in a proper parabolic subsystem of the root system of $\mathbf{L}^*$. But in a given root system there is only a finite number of subsystems, whence only a finite number of $\mathbf{G}^*$-conjugacy classes of possible groups $C_{\mathbf{G}^*}(s)$, depending only on the type of the root system of $\mathbf{G}$. Any one of these groups is the centralizer of a finite number of elements of the corresponding $\mathbf{L}^{*\prime}$ by the following lemma applied to $\mathbf{H} = C_{\mathbf{G}^*}(s) \cap \mathbf{L}^{*\prime}$ and $\mathbf{G} = \mathbf{L}^{*\prime}$.

14.11 LEMMA. *Let* $\mathbf{G}$ *be a connected reductive group and let* $\mathbf{H}$ *be a connected subgroup of maximal rank of* $\mathbf{G}$. *Assume that the root system of* $\mathbf{H}$ *(relative to some torus* $\mathbf{T}$*) is not contained in any parabolic subsystem of the root system of* $\mathbf{G}$ *relative to* $\mathbf{T}$*; then* $Z(\mathbf{H})^\circ = Z(\mathbf{G})^\circ$. *In particular, if* $\mathbf{G}$ *is semi-simple, then so is* $\mathbf{H}$.

PROOF: The centralizer in $\mathbf{G}$ of $Z(\mathbf{H})^\circ$ is a Levi subgroup of $\mathbf{G}$ containing $\mathbf{H}$ (see 1.22), so is equal to $\mathbf{G}$ and we have $Z(\mathbf{H})^\circ \subset Z(\mathbf{G})^\circ$. As $\mathbf{H}$ is of maximal rank it contains $Z(\mathbf{G})^\circ$, whence the result. ■

So we get a finite number of $\mathbf{G}^*$-conjugacy classes of possible elements s. As the centralizers of these elements are connected, each $\mathbf{G}^*$-conjugacy class gives at most one rational conjugacy class, so there is a bound for the cardinality of $\mathcal{S}(\mathbf{G}^*)$ that depends only on the type of the root system of $\mathbf{G}^*$.

We shall be done if we bound for each s the cardinality of $\mathcal{E}(\mathbf{G}^F, (s))$. Each Deligne-Lusztig character has at most $|W|$ components by 11.15, and the number of Deligne-Lusztig characters "in" a Lusztig series $\mathcal{E}(\mathbf{G}^F, (s))$ is the number of classes of maximal rational tori in the centralizer of s, so is again bounded by $|W|$.

We now recall a well-known result from algebraic geometry which will be used in the proofs of 14.13 and of 14.20.

14.12 LEMMA. *The dimension of any fibre of a surjective morphism from an irreducible variety* $\mathbf{X}$ *to another* $\mathbf{Y}$ *is at least* $\dim(\mathbf{X})-\dim(\mathbf{Y})$; *moreover this inequality is an equality for an open dense subset of* $\mathbf{Y}$.

PROOF: For the proof of the inequality see [Ha, II, exercise 3.22]. The second statement is proved in [St2, appendix to 2.11, prop. 2]. ∎

We can now prove

14.13 THEOREM. *Regular unipotent elements exist.*

PROOF: The variety $\mathbf{G}_u$ of unipotent elements is a finite union of conjugacy classes, so has the same dimension as one of them. Thus the existence of a regular unipotent class is equivalent to the equality $\operatorname{codim}(\mathbf{G}_u) = \operatorname{rk}(\mathbf{G})$. Note that we have $\operatorname{codim}(\mathbf{G}_u) \geq \operatorname{rk}(\mathbf{G})$, as the codimension of any conjugacy class is at most $\operatorname{rk}(\mathbf{G})$ by 14.2 (i). To compute the dimension of $\mathbf{G}_u$, we consider the unipotent radical $\mathbf{U}$ of a Borel subgroup $\mathbf{B} = \mathbf{TU}$ and the morphism $\mathbf{G}\times\mathbf{U} \to \mathbf{G}$ given by $(g,u) \mapsto {}^g u$. Its image is $\mathbf{G}_u$. Consider an element $u = \prod_\alpha u_\alpha$ (where α runs over the roots of $\mathbf{G}$ relative to $\mathbf{T}$; see 0.31 (v)) such that $u_\alpha \neq 1$ for any simple α. We shall see in the proof of "(iv) implies (iii)" in 14.14 that u is contained in only one Borel subgroup. So (g,v) is in the fibre of u if and only if $g \in \mathbf{B}$ and $v = {}^{g^{-1}}u$; thus the fibre of u is isomorphic to $\mathbf{B}$. By 14.12 we have $\dim(\mathbf{B}) \geq \dim(\mathbf{G}) + \dim(\mathbf{U}) - \dim(\mathbf{G}_u)$. As $\dim(\mathbf{B}) - \dim(\mathbf{U}) = \dim(\mathbf{T}) = \operatorname{rk}(\mathbf{G})$, we get $\operatorname{codim}(\mathbf{G}_u) \leq \operatorname{rk}(\mathbf{G})$, whence the result. ∎

Once we know they exist we can prove the following characterizations of regular unipotent elements.

14.14 PROPOSITION. *Let* u *be a unipotent element of* $\mathbf{G}$; *the following properties are equivalent:*

(i) *u is regular.*
(ii) *There is only a finite number of Borel subgroups containing u.*
(iii) *There is only one Borel subgroup containing u.*
(iv) *There exists a Borel subgroup $\mathbf{B}$ containing u and a maximal torus $\mathbf{T}$ of $\mathbf{B}$ such that in the decomposition $u = \prod_\alpha u_\alpha$, where α runs over the roots of $\mathbf{G}$ relative to $\mathbf{T}$ (see 0.31 (v)), none of the u_α for α simple is equal to 1.*
(v) *Property (iv) is true for any Borel subgroup $\mathbf{B}$ containing u and any maximal torus of $\mathbf{B}$.*

PROOF: It is obvious that (iii) implies (ii) and that (v) implies (iv). We prove that (ii) implies (v). Let $\mathbf{B}$ be a Borel subgroup containing u and $\mathbf{T}$ be a maximal torus of $\mathbf{B}$; if one of the u_α is 1, then for any $v \in \mathbf{U}_\alpha$ the element ${}^v u$ has also a trivial component in $\mathbf{U}_\alpha$ by the commutation formulae 0.31 (iv). Let $n \in N_{\mathbf{G}}(\mathbf{T})$ be a representative of the simple reflection s_α of $W(\mathbf{T})$ corresponding to α; then ${}^{nv}u$ is in $\mathbf{B}$ since the only root subgroup of $\mathbf{B}$ which is mapped onto a negative root subgroup under the action of n is $\mathbf{U}_\alpha$. Whence we see that u is in ${}^{vs_\alpha}\mathbf{B}$ for any $v \in \mathbf{U}_\alpha$. These Borel subgroups are all distinct since two distinct elements of $\mathbf{U}_\alpha$ differ by a non-trivial element of $\mathbf{U}_\alpha$, and so by an element which is not in ${}^{s_\alpha}\mathbf{B} = N_{\mathbf{G}}({}^{s_\alpha}\mathbf{B})$. Thus we have got an infinite number (at least an affine line) of Borel subgroups containing u.

We show now that (iv) implies (iii). We denote by $\mathbf{B}$ (resp. $\mathbf{T}$) the Borel subgroup (resp. maximal torus) given by (iv). Let $\mathbf{B}_1$ be a Borel subgroup containing u and let $\mathbf{T}_1$ be a torus contained in $\mathbf{B} \cap \mathbf{B}_1$ (see 1.5); then $\mathbf{T} = {}^v\mathbf{T}_1$ with $v \in \mathbf{U}$, so that ${}^v\mathbf{B}_1$ is a Borel subgroup containing $\mathbf{T}$ and so equals ${}^w\mathbf{B}$ for some $w \in W(\mathbf{T})$. Thus we see that ${}^w\mathbf{B}$ contains ${}^v u$. By the commutation formulae 0.31 (iv), the element ${}^v u$ satisfies the same property as u, but if α is a simple root such that ${}^{w^{-1}}\alpha$ is negative, then the root subgroup $\mathbf{U}_\alpha$ has a trivial intersection with ${}^w\mathbf{B}$, so the component of ${}^v u$ in $\mathbf{U}_\alpha$ is necessarily trivial. This is a contradiction if w is not 1, *i.e.*, if $\mathbf{B}_1 \neq \mathbf{B}$.

So far we have proved the equivalence of (ii), (iii), (iv), (v). We shall show now that (i) is equivalent to these four properties. First we prove that (i) implies (iv). Let u be a regular unipotent element and $\mathbf{B} = \mathbf{TU}$ be a Borel subgroup containing u; we write $u = \prod_\alpha u_\alpha$ with $u_\alpha \in \mathbf{U}_\alpha$. Assume that for some simple α we have $u_\alpha = 1$. The element u is in the parabolic subgroup $\mathbf{P} = \mathbf{B} \cup \mathbf{B}s_\alpha\mathbf{B}$ where s_α is the reflection with respect to α. In that parabolic subgroup the conjugacy class of u contains only elements of $\mathbf{U}$ whose component in $\mathbf{U}_\alpha$ is 1, so is of dimension at most $\dim(\mathbf{U}) - 1$. Hence the dimension of the centralizer of u in $\mathbf{P}$ is at least $\dim(\mathbf{P}) - (\dim(\mathbf{U}) - 1)$. As $\dim(\mathbf{P}) = \dim(\mathbf{U}) + \dim(\mathbf{T}) + 1$, we get $\dim(C_{\mathbf{P}}(u)) \geq \mathrm{rk}(\mathbf{G}) + 2$, which is a contradiction by 14.2.

We now prove that any u satisfying (i) is conjugate to any v satisfying (iv), which will finish the proof and prove in addition that all regular elements are conjugate. We already know that u satisfies (iv). Replacing v by a conjugate, we may assume that the groups $\mathbf{B}$ and $\mathbf{T}$ of property (iv) are the same for u and for v. Replacing v by a $\mathbf{T}$-conjugate, we may assume that the components of v and u in $\mathbf{U}_\alpha$ for α simple are equal (see 0.31 (i) and 0.22), so that vu^{-1} is in $\mathbf{U}^* = \prod_{\alpha\in\Phi^+-\Pi}\mathbf{U}_\alpha$. As the orbits of a unipotent group acting on a variety are closed (see [St2, 2.5, proposition]), the set $\{[x,u] \mid x \in \mathbf{U}\}$ is a closed subvariety of $\mathbf{U}^*$. Its codimension in $\mathbf{U}$ is $\dim(C_{\mathbf{U}}(u))$. If we prove that this codimension is equal to that of $\mathbf{U}^*$, we shall get the equality $\mathbf{U}^* = \{[x,u] \mid x \in \mathbf{U}\}$, whence the existence of an $x \in \mathbf{U}$ such that $[x,u] = vu^{-1}$, so $v = {}^x u$, whence the proposition. But the codimension of $\mathbf{U}^*$ in $\mathbf{U}$ is the semi-simple rank of $\mathbf{G}$, equal to $\mathrm{rk}(\mathbf{G}) - \dim(Z(\mathbf{G}))$, and this is also the dimension of $C_{\mathbf{U}}(u)$ by the following lemma.

14.15 LEMMA. *Let u be a regular unipotent element and let $\mathbf{U}$ be the unipotent radical of the unique Borel subgroup containing u; then $C_{\mathbf{G}}(u) = Z(\mathbf{G}).C_{\mathbf{U}}(u)$.*

PROOF: If g centralizes u then u is in ${}^{g^{-1}}\mathbf{B}$, so, as (i) implies (iii) in 14.14, we get that g normalizes $\mathbf{B}$, *i.e.*, is in $\mathbf{B}$. So the Jordan decomposition of g is of the form $g = tv$ with t semi-simple in $C_{\mathbf{B}}(u)$ and v unipotent. By 14.14 (v) as t is in some maximal torus of $\mathbf{B}$, we get $t \in Z(\mathbf{G})$, whence the lemma. ∎

Note that in the proof of 14.14 we have shown

14.16 PROPOSITION. *All regular unipotent elements are conjugate in* $\mathbf{G}$.

We have also proved that any element of $\mathbf{U}^*$ is a commutator of a regular element and another element of $\mathbf{U}$, so in particular

14.17 PROPOSITION. *The derived group of* $\mathbf{U}$ *is* $\mathbf{U}^*$.

The following important result involves the notion of **good characteristic** for $\mathbf{G}$. The characteristic is good if it does not divide the coefficients of the highest root of the root system associated to $\mathbf{G}$. Bad (*i.e.*, not good) characteristics are $p = 2$ if the root system is of type B_n, C_n or D_n, $p = 2, 3$ in types G_2, F_4, E_6, E_7 and $p = 2, 3, 5$ in type E_8 (see [Bbk, VI §4]).

14.18 PROPOSITION. *Let u be a regular unipotent element of $\mathbf{G}$; if the characteristic is good for $\mathbf{G}$ then $C_{\mathbf{U}}(u)$ is connected, otherwise u is not in $C^\circ_{\mathbf{U}}(u)$, and $C_{\mathbf{U}}(u)/C^\circ_{\mathbf{U}}(u)$ is cyclic generated by the image of u.*

PROOF: See [T. A. SPRINGER Some arithmetic results on semi-simple Lie algebras, *Publications Mathématiques de l'IHES*, **30** (1966) 115–141, 4.11, 4.12] and [B. LOU The centralizer of a regular unipotent element in a semi-simple algebraic group, *Bull. AMS*, **74** 1144–1146]. Most of the proof is by a case-by-case check. ∎

From the above results on unipotent and semi-simple regular elements, we can deduce results for all regular elements.

14.19 COROLLARY. *An element x is regular if and only if it is contained in a finite number of Borel subgroups, and this number is equal to $|W/W^\circ(s)|$ where s is the semi-simple part of x.*

PROOF: By 14.4 if su is the Jordan decomposition of x, then x is regular if and only if u is regular in $C^\circ_{\mathbf{G}}(s)$, which is equivalent to u being contained in a finite number of Borel subgroups (in fact one) of $C^\circ_{\mathbf{G}}(s)$. But a Borel subgroup of $C^\circ_{\mathbf{G}}(s)$ is of the form $\mathbf{B} \cap C^\circ_{\mathbf{G}}(s)$ where $\mathbf{B}$ is a Borel subgroup of $\mathbf{G}$ containing a maximal torus of $C^\circ_{\mathbf{G}}(s)$ (see 2.2 (i)). As the number of Borel subgroups containing a maximal torus in a connected reductive group is finite, we get the first assertion. This number is equal to the cardinality of the Weyl group, so is $|W|$ in $\mathbf{G}$ and $|W^\circ(s)|$ in $C^\circ_{\mathbf{G}}(s)$; moreover any two Borel subgroups of $C^\circ_{\mathbf{G}}(s)$ are contained in the same number of Borel subgroups of $\mathbf{G}$ since they are conjugate, whence the second assertion. ■

We now give a general result about centralizers. This result is due to Springer [A note on centralizers in semi-simple groups, *Indagationes Math.* **28** (1966), 75–77]. We follow here the proof given in [St2, remark of 3.5].

14.20 PROPOSITION. *The centralizer of any element contains an abelian subgroup of dimension at least* rk($\mathbf{G}$).

PROOF: The result is clear for semi-simple elements. The idea of the proof is to use the fact that semi-simple elements are dense in $\mathbf{G}$ (see 14.9). Consider the variety

$$\mathbf{S}_n = \{ (x_1, x_2, \ldots, x_n) \in \mathbf{G}^n \mid \text{all } x_i \text{ belong to the same torus} \};$$

then the projection $(x_1, x_2, \ldots, x_n) \mapsto (x_1, x_2, \ldots, x_{n-1})$ is surjective from $\mathbf{S}_n$ to $\mathbf{S}_{n-1}$. Moreover, as $\mathbf{S}_n$ is the image of $\mathbf{G} \times \mathbf{T}^n$ by the map $(g, t_1, \ldots, t_n) \mapsto ({}^g t_1, \ldots, {}^g t_n)$, it is irreducible, so its closure $\overline{\mathbf{S}}_n$ is also irreducible. It follows that the projection is also surjective from $\overline{\mathbf{S}}_n$ to $\overline{\mathbf{S}}_{n-1}$. Note also that the components of an element of $\mathbf{S}_n$ commute with each other, so the same property is true for $\overline{\mathbf{S}}_n$. By the density property of semi-simple elements we have $\overline{\mathbf{S}}_1 = \mathbf{G}$. Now let x be an arbitrary element of $\mathbf{G}$ and choose n and $(x_1, x_2, \ldots, x_{n-1}) \in \mathbf{G}^{n-1}$ such that $(x, x_1, \ldots, x_{n-1}) \in \overline{\mathbf{S}}_n$ and that $C_{\mathbf{G}}(x, x_1, \ldots, x_{n-1})$ is minimal (to get the existence of such an n and such an $(n-1)$-tuple, argue first on the dimension of the centralizer, then on the cardinality of $C_{\mathbf{G}}(x, x_1, \ldots, x_{n-1})/C_{\mathbf{G}}(x, x_1, \ldots, x_{n-1})^\circ)$. Then any $z \in \mathbf{G}$ such that $(x, x_1, \ldots, x_{n-1}, z) \in \overline{\mathbf{S}}_{n+1}$, satisfies $C_{\mathbf{G}}(x, x_1, \ldots, x_{n-1}, z) = C_{\mathbf{G}}(x, x_1, \ldots, x_{n-1})$, so is in the centre of $C_{\mathbf{G}}(x, x_1, \ldots, x_{n-1})$. The set of such elements z is thus an abelian subgroup of $C_{\mathbf{G}}(x)$, so we shall be done if we prove that the fibres of the projection $\overline{\mathbf{S}}_{n+1} \to \overline{\mathbf{S}}_n$ are of dimension at least rk($\mathbf{G}$). By 14.12 (which we can use since the morphism is surjective and $\overline{\mathbf{S}}_{n+1}$

is irreducible) the dimension of any fibre is at least $\dim(\overline{\mathbf{S}}_{n+1}) - \dim(\overline{\mathbf{S}}_n)$, and this inequality is an equality for a dense open subset of $\overline{\mathbf{S}}_n$. But the fibre of an element of $\mathbf{S}_n$ contains at least a maximal torus, so has dimension at least $\mathrm{rk}(\mathbf{G})$ whence $\dim(\overline{\mathbf{S}}_{n+1}) - \dim(\overline{\mathbf{S}}_n) \geq \mathrm{rk}(\mathbf{G})$, which gives the result. ■

14.21 COROLLARY. *If x is regular then $C^\circ_{\mathbf{G}}(x)$ is abelian.*

PROOF: We know that the dimension of $C^\circ_{\mathbf{G}}(x)$ is equal to $\mathrm{rk}(\mathbf{G})$, but by 14.20, $C^\circ_{\mathbf{G}}(x)$ contains an abelian subgroup of that dimension, whence the result. ■

We now study rational classes of regular unipotent elements. We assume that $\mathbf{G}$ is defined over $\mathbb{F}_q$ with Frobenius endomorphism F.

14.22 REMARK. Note that all the rational classes of regular unipotent elements have the same cardinality: indeed the centralizer $C_{\mathbf{G}}(u)$ of such an element u is abelian by 14.18 and 14.21 and an easy computation shows that if $x \in \mathbf{G}$ is such that $\mathcal{L}(x) \in C_{\mathbf{G}}(u)$ the conjugation by x maps $C_{\mathbf{G}^F}(u)$ onto $C_{\mathbf{G}^F}({}^x u)$, so that these two centralizers have the same cardinality.

We shall use the notation $H^1(F, \mathbf{H})$ for the set of F-conjugacy classes of an algebraic group $\mathbf{H}$ defined over $\mathbb{F}_q$. Recall that by 3.22 we have $H^1(F, \mathbf{H}) = H^1(F, \mathbf{H}/\mathbf{H}^\circ)$. Note also that if $\mathbf{H}$ is an abelian group the F-conjugacy classes are the cosets with respect to the subgroup $\mathcal{L}(\mathbf{H})$ (the Lang map is a group morphism in this case), so $H^1(F, \mathbf{H}) \simeq \mathbf{H}/\mathcal{L}(\mathbf{H})$ and, as the kernel of $\mathcal{L}$ is $\mathbf{H}^F$ we have $|H^1(F, \mathbf{H})| = |\mathbf{H}^F|$.

14.23 PROPOSITION. *The number of regular unipotent elements in $\mathbf{G}^F$ is $|\mathbf{G}^F|/(|Z(\mathbf{G})^{\circ F}|q^l)$ where l is the semi-simple rank of $\mathbf{G}$.*

PROOF: A regular unipotent element lies in only one Borel subgroup, so the number of rational regular unipotent elements is equal to the product of the number of rational Borel subgroups (which is $|\mathbf{G}^F/\mathbf{B}^F|$ for any rational Borel subgroup $\mathbf{B}$) with the number of such elements in a given Borel subgroup. By 14.14 (v), the number of rational regular unipotent elements in a Borel subgroup $\mathbf{B} = \mathbf{T}\mathbf{U}$ is $|\mathbf{U}^{*F}|$ times the number of elements of $\mathbf{U}^F/\mathbf{U}^{*F}$ whose every component is non-trivial in the decomposition $\mathbf{U}/\mathbf{U}^* = \prod \mathbf{U}_\alpha$ (we shall say that these elements are regular in $\mathbf{U}/\mathbf{U}^*$). By 0.31 (i) and 0.22 any two regular elements of $\mathbf{U}/\mathbf{U}^*$ are conjugate by an element of $\mathbf{T}$, but the centralizer of a regular element in $\mathbf{T}$ is $Z(\mathbf{G})$, so by 3.21 the number of $\mathbf{T}^F$-orbits of regular rational elements of $\mathbf{U}^F/\mathbf{U}^{*F}$ is $|H^1(F, Z(\mathbf{G}))|$. In each such orbit there are $|\mathbf{T}^F/Z(\mathbf{G})^F|$ elements, so the number of regular elements in $\mathbf{G}^F$ is $|\mathbf{G}^F|/(|\mathbf{T}^F||\mathbf{U}^F|)|\mathbf{U}^{*F}||H^1(F, Z(\mathbf{G}))||\mathbf{T}^F|/|Z(\mathbf{G})^F|$. Now $\mathbf{U}^F/\mathbf{U}^{*F}$ has cardinality q^l (by, *e.g.*, 10.11 (ii), as it is an affine space of dimension *l*) and $H^1(F, Z(\mathbf{G})) = H^1(F, Z(\mathbf{G})/Z(\mathbf{G})^\circ)$ has the same cardinality as $Z(\mathbf{G})^F/Z(\mathbf{G})^{\circ F}$ because $Z(\mathbf{G})/Z(\mathbf{G})^\circ$ is abelian, so the result follows. ∎

Note that by 3.21 and the above proof $\mathcal{L}_{\mathbf{T}}^{-1}(Z(\mathbf{G}))$ permutes transitively the regular elements of $\mathbf{U}^F/\mathbf{U}^{*F}$.

14.24 PROPOSITION. *If the characteristic is good for* $\mathbf{G}$*, the* $\mathbf{G}^F$*-conjugacy classes of rational regular unipotent elements are parametrized by the* F*-conjugacy classes of* $Z(\mathbf{G})/Z(\mathbf{G})^\circ$.

PROOF: First note that, as all regular unipotent elements are conjugate in $\mathbf{G}$, rational regular unipotent elements exist (see 3.12). Then 3.25 tells us that the $\mathbf{G}^F$-conjugacy classes of rational regular unipotent elements are parametrized by the F-conjugacy classes of $C_{\mathbf{G}}(u)/C^\circ_{\mathbf{G}}(u)$, if u is such an element. If the characteristic is good, by 14.18, we have $C_{\mathbf{G}}(u)/C^\circ_{\mathbf{G}}(u) = Z(\mathbf{G})/Z(\mathbf{G})^\circ$, whence the result. ∎

The following corollary is clear.

14.25 COROLLARY. *If the characteristic is good and the centre of* $\mathbf{G}$ *is connected, there is only one class of regular unipotent elements in* $\mathbf{G}^F$.

Note that the parametrization of 14.24 is well-defined only once we have chosen a rational class of regular unipotent elements. From now on we shall fix such an element which will be denoted by u_1.

When the characteristic is bad $C_{\mathbf{G}}(u_1)$ is not connected. The rational conjugacy classes of regular unipotent elements are parametrized by

$$H^1(F, Z(\mathbf{G})/Z(\mathbf{G})^\circ \times C_{\mathbf{G}}(u_1)/C^\circ_{\mathbf{G}}(u_1));$$

the first projection maps these F-classes onto the F-classes of $Z(\mathbf{G})/Z(\mathbf{G})^\circ$. We shall denote by $\mathcal{U}_z$ the set of regular unipotent elements of $\mathbf{G}^F$ in a class parametrized by any element of the fibre of $z \in H^1(F, Z(\mathbf{G})/Z(\mathbf{G})^\circ)$. If $\mathbf{U}$ is the unipotent radical of some Borel subgroup and if $\mathbf{T}$ is a maximal torus of $\mathbf{B}$ we put $\mathbf{U}^* = \prod_{\alpha\in\Phi^+-\Pi} \mathbf{U}_\alpha$, where Φ is the root system of $\mathbf{G}$ relative to $\mathbf{T}$ and Π the basis of Φ defined by $\mathbf{B}$. The sets $\mathcal{U}_z$ are characterized by the following property.

14.26 PROPOSITION. *With the above notation (and assuming* $\mathbf{B}$ *and* $\mathbf{T}$ *rational), two regular elements* u *and* u' *of* $\mathbf{U}^F$ *are in the same set* $\mathcal{U}_z$ *if and only if there exists* $t \in \mathbf{T}^F$ *such that* $u^{-1}\,{}^t u'$ *is in* $\mathbf{U}^*$.

PROOF: There is a rational conjugate of u_1 in $\mathbf{U}$ which we denote again by u_1. Any element which conjugates u_1 to u is in $\mathbf{B}$ by 14.14 (iii). So we have

$u = {}^{sv}u_1$ with $s \in \mathbf{T}$ and $v \in \mathbf{U}$. The class of u is parametrized by the F-class of $v^{-1}s^{-1F}s^F v$ in $H^1(F, Z(\mathbf{G})/Z(\mathbf{G})^\circ \times C_{\mathbf{G}}(u_1)/C^\circ_{\mathbf{G}}(u_1))$ (we know by 14.15 that $s^{-1F}s$ is in $Z(\mathbf{G})$). The element u' is in the same $\mathcal{U}_z$ if and only if it can be written ${}^{s'v'}u_1$ with $s^{-1F}s$ and $s'^{-1F}s'$ in the same F-class of $Z(\mathbf{G})$, which means that we have ${}^{ts'v'}u_1 = {}^{sv'}u_1$ with $t \in \mathbf{T}^F$. So u and ${}^tu'$ are conjugate under $\mathbf{U}$, which implies by 0.31 (iv) that they have the same image in $\mathbf{U}/\mathbf{U}^*$. Conversely, if ${}^{sv}u_1$ and ${}^{ts'v'}u_1$ have the same image in $\mathbf{U}/\mathbf{U}^*$, the element $s^{-1}ts'$ must be central by 14.14 (v) and 0.31 (i), so $s^{-1F}s$ and $s'^{-1F}s'$ are in the same F-class of $Z(\mathbf{G})$, whence the result. ∎

We shall now give the definition of the Gelfand-Graev representations. In the following we fix a rational Borel subgroup $\mathbf{B}$ of $\mathbf{G}$ and a rational maximal torus $\mathbf{T}$ of $\mathbf{B}$, so that we have the Levi decomposition $\mathbf{B} = \mathbf{TU}$. We denote by Π a basis of the root system of $\mathbf{G}$ relative to $\mathbf{T}$. The Frobenius endomorphism defines a permutation τ of Π (see chapter 8). For any orbit $\mathcal{O}$ of τ in Π, we denote by $\mathbf{U}_{\mathcal{O}}$ the image of $\prod_{\alpha\in\mathcal{O}} \mathbf{U}_\alpha$ in $\mathbf{U}/\mathbf{U}^*$. It is a commutative group isomorphic to $\mathbb{G}_a^{|\mathcal{O}|}$. The action of F maps $\mathbf{U}_\alpha$ on $\mathbf{U}_{\tau\alpha}$ for each $\alpha \in \Pi$ (and is an isomorphism of abstract groups). The action of $F^{|\mathcal{O}|}$ stabilizes each $\mathbf{U}_\alpha$ for $\alpha \in \mathcal{O}$, so the group of rational points $\mathbf{U}_{\mathcal{O}}^F$ is isomorphic to $\mathbf{U}_\alpha^{F^{|\mathcal{O}|}}$ for any $\alpha \in \mathcal{O}$. We choose such an α: the group $\mathbf{U}_\alpha$ is isomorphic to the additive group and we can choose this isomorphism so that the image in $\mathbf{U}_\alpha$ of $1 \in \mathbb{G}_a$ is fixed by $F^{|\mathcal{O}|}$ (by 0.44 (i) and the Lang-Steinberg theorem); we then get $\mathbf{U}_{\mathcal{O}}^F \simeq \mathbb{F}_{q^{|\mathcal{O}|}}$ (see 3.8). In the following we shall assume that we have made such choices for each orbit $\mathcal{O}$.

14.27 DEFINITION. *A linear character $\phi \in \mathrm{Irr}(\mathbf{U}^F)$ is* **regular** *if:*
(i) Its restriction to $\mathbf{U}^{*F}$ is trivial,
(ii) Its restriction to $\mathbf{U}_{\mathcal{O}}^F$ is not trivial for any orbit $\mathcal{O}$ of τ in Π.

Note that, as $\mathbf{U}/\mathbf{U}^*$ is abelian (isomorphic to a product of additive groups), the group $\mathbf{U}^F/\mathbf{U}^{*F} \simeq (\mathbf{U}/\mathbf{U}^*)^F$ is abelian too, so $\mathbf{U}^{*F}$ contains the derived group of $\mathbf{U}^F$ (actually it is almost always equal to that derived group, the only exceptions for quasi-simple groups being groups of type B_2 or F_4 over $\mathbb{F}_2$ and groups of type G_2 over $\mathbb{F}_3$; see [R. B. HOWLETT On the degrees of Steinberg characters of Chevalley groups *Math. Zeitschrift,* **135** (1974), 125–135, lemma 7] where the proof is given in the case of split groups, for non-split groups the proof is similar). Note also that the quotient map is an isomorphism of varieties $\prod_{\alpha\in\Pi} \mathbf{U}_\alpha \xrightarrow{\sim} \mathbf{U}/\mathbf{U}^*$ compatible with the actions of $\mathbf{T}$ and that this isomorphism induces a bijection on rational elements. We thus have an action of $\mathbf{T}^F$ on regular characters.

14.28 PROPOSITION. *The $\mathbf{T}^F$-orbits of regular characters of $\mathbf{U}^F$ are in one-to-one correspondence with $H^1(F, \mathbf{Z}_{\mathbf{G}})$.*

PROOF: By the remark after 14.23 $\mathcal{L}_{\mathbf{T}}^{-1}(Z(\mathbf{G}))$ acts transitively on regular elements of $\mathbf{U}^F/\mathbf{U}^{*F}$. By the remarks before 14.27, a regular character is identified with an element of $\prod_{\mathcal{O}\in\Pi/\tau}\{\mathrm{Irr}(\mathbb{F}^+_{q^{|\mathcal{O}|}})-\{\mathrm{Id}\}\}$ and a regular element of $\mathbf{U}^F/\mathbf{U}^{*F}$ is identified with an element of $\prod_{\mathcal{O}\in\Pi/\tau}\{\mathbb{F}_{q^{|\mathcal{O}|}}-\{0\}\}$. So we see that $\mathcal{L}_{\mathbf{T}}^{-1}(Z(\mathbf{G}))$ acts transitively on the set of regular characters of $\mathbf{U}^F$. It is also clear that the stabilizer of a regular character is $Z(\mathbf{G})$. So, by 3.25 we get the result (in 3.25 the group is assumed to be connected but the only property actually needed is that the stabilizer of an element is contained in the image of the Lang map, so we can apply 3.25 to $\mathcal{L}_{\mathbf{T}}^{-1}(Z(\mathbf{G}))$ in our case). ∎

Note that, by the above proof, the regular characters of $\mathbf{U}^F$ are parametrized by $\mathcal{L}_{\mathbf{T}}^{-1}(Z(\mathbf{G}))/Z(\mathbf{G})$. As in the case of regular unipotent elements, this parametrization is well-defined once we have chosen a regular character. From now on we shall fix such a character which will be denoted by ψ_1. We make the choice in the following way: we fix an integer N, multiple of $|\mathcal{O}|$ for any orbit $\mathcal{O}$, and choose a character χ of $\mathbb{F}_{q^N}$ having a non-trivial restriction to $\mathbb{F}_q$; then we take for ψ_1 the product of the restrictions of χ to the additive groups $\mathbb{F}_{q^{|\mathcal{O}|}}$. The orbit of regular characters parametrized by $z\in H^1(F,Z(\mathbf{G}))$ will be denoted by Ψ_z. If $\dot z$ is the image in $\mathcal{L}_{\mathbf{T}}^{-1}(Z(\mathbf{G}))/Z(\mathbf{G})$ of an element $t\in\mathcal{L}_{\mathbf{T}}^{-1}(Z(\mathbf{G}))$ such that the F-conjugacy class of $\mathcal{L}(t)$ is equal to z, the character $\psi_{\dot z}={}^t\psi_1$ is in Ψ_z.

14.29 DEFINITION. *For $z\in Z(\mathbf{G})$ we define the* **Gelfand-Graev representation** Γ_z *(or $\Gamma_z^{\mathbf{G}}$) by $\Gamma_z=\mathrm{Ind}_{\mathbf{U}^F}^{\mathbf{G}^F}(\psi_{\dot z})$.*

The following result is due to Steinberg:

14.30 THEOREM. *The Gelfand-Graev representations are multiplicity free.*

PROOF: We shall not give the proof here. It may be found, *e.g.*, in [Ca, 8.1.3]; the proof given there is written under the assumption that $Z(\mathbf{G})$ is connected, but it applies without any change to our case. ∎

We now study the effect of Harish-Chandra restriction and of duality on the Gelfand-Graev representations. For this we need some notation and preliminary results. If $\mathbf{L}$ is a rational Levi subgroup containing $\mathbf{T}$ of a rational parabolic subgroup of $\mathbf{G}$, to parametrize the regular characters of $\mathbf{U}^F\cap\mathbf{L}^F$ by $H^1(F,Z(\mathbf{L}))$ we have to fix a particular regular character: we choose $\mathrm{Res}_{\mathbf{U}^F\cap\mathbf{L}^F}^{\mathbf{U}^F}\psi_1$. The next lemma will allow us to compare the parametrizations in $\mathbf{G}$ and in $\mathbf{L}$.

14.31 LEMMA. *The inclusion $Z(\mathbf{G})\subset Z(\mathbf{L})$ induces a surjective map*

$$h_{\mathbf{L}}: H^1(F,Z(\mathbf{G}))\to H^1(F,Z(\mathbf{L})).$$

PROOF: First note that $Z(\mathbf{L}) \supset Z(\mathbf{G})$ and that the quotient $Z(\mathbf{L})/Z(\mathbf{G})$ is connected, by 0.36 and 13.14 (ii), as it is the centre of the Levi subgroup $\mathbf{L}/Z(\mathbf{G})$ of the group $\mathbf{G}/Z(\mathbf{G})$ (use a similar argument to that in 0.36 in $\mathbf{L}$). So any element of $Z(\mathbf{L})$ is in the image of the Lang map modulo $Z(\mathbf{G})$, which means that it is F-conjugate to an element of $Z(\mathbf{G})$ (it could also be said that we have a long exact sequence of Galois cohomology groups

$$\ldots \to H^1(F, Z(\mathbf{G})) \to H^1(F, Z(\mathbf{L})) \to 1,$$

as $H^1(F, Z(\mathbf{L})/Z(\mathbf{G}))$ is trivial). ∎

14.32 PROPOSITION. *With the above notation we have*

$$^*R^{\mathbf{G}}_{\mathbf{L}}(\Gamma^{\mathbf{G}}_z) = \Gamma^{\mathbf{L}}_{h_{\mathbf{L}}(z)}.$$

PROOF: We have to show that, for any class function f on $\mathbf{L}^F$, we have

$$\langle R^{\mathbf{G}}_{\mathbf{L}}(f), \Gamma^{\mathbf{G}}_z \rangle_{\mathbf{G}^F} = \langle f, \Gamma^{\mathbf{L}}_{h_{\mathbf{L}}(z)} \rangle_{\mathbf{L}^F}.$$

By the definition of Harish-Chandra induction we have $R^{\mathbf{G}}_{\mathbf{L}}(f) = \operatorname{Ind}^{\mathbf{G}^F}_{\mathbf{P}^F} \tilde{f}$ where $\mathbf{P} = \mathbf{L}.\mathbf{U}$ is the parabolic subgroup which contains $\mathbf{B}$ and has $\mathbf{L}$ as a Levi subgroup, and where $\tilde{f}$ is the function on $\mathbf{P}^F$ defined by $\tilde{f}(lu) = f(l)$. So the left-hand side is equal to

$$\langle \operatorname{Ind}^{\mathbf{G}^F}_{\mathbf{P}^F} \tilde{f}, \operatorname{Ind}^{\mathbf{G}^F}_{\mathbf{U}^F} \psi_{\dot{z}} \rangle_{\mathbf{G}^F} = \sum_{g \in \mathbf{P}^F\backslash\mathbf{G}^F/\mathbf{U}^F} \langle \tilde{f}, {}^g\psi_{\dot{z}} \rangle_{\mathbf{P}^F \cap {}^g\mathbf{U}^F},$$

the last equality by the ordinary Mackey formula. We know (see 5.5) that there is a cross section of $\mathbf{P}\backslash\mathbf{G}/\mathbf{B}$ consisting of the I-reduced elements of $W = W(\mathbf{T})$ (if $\mathbf{L}$ is the standard Levi subgroup $\mathbf{L}_I$). So we can choose as a cross section of $\mathbf{P}^F\backslash\mathbf{G}^F/\mathbf{U}^F$ a set of representatives in $N_{\mathbf{G}}(\mathbf{T})^F$ of I-reduced elements of W^F (see 5.6 (ii)). If n is such an element (a representative of $w \in W^F$) we have

$$\mathbf{P}^F \cap {}^n\mathbf{U}^F = (\mathbf{L}^F \cap {}^n\mathbf{U}^F).(\mathbf{V}^F \cap {}^n\mathbf{U}^F),$$

where $\mathbf{V}$ is the unipotent radical of $\mathbf{P}$, and the restriction of $\tilde{f}$ to $\mathbf{P}^F \cap {}^n\mathbf{U}^F$ is equal to $\operatorname{Res}^{\mathbf{L}^F}_{\mathbf{L}^F \cap {}^n\mathbf{U}^F}(f) \times \operatorname{Id}$ in that decomposition. So we have

$$\langle \tilde{f}, {}^n\psi_{\dot{z}} \rangle_{\mathbf{P}^F \cap {}^n\mathbf{U}^F} = \langle \operatorname{Res}^{\mathbf{L}^F}_{\mathbf{L}^F \cap {}^n\mathbf{U}^F} f, {}^n\psi_{\dot{z}} \rangle_{\mathbf{L}^F \cap {}^n\mathbf{U}^F} \langle \operatorname{Id}, \psi_{\dot{z}} \rangle_{{}^{n^{-1}}\mathbf{V}^F \cap \mathbf{U}^F}.$$

But, by the definition of $\psi_{\dot{z}}$, the scalar product $\langle \operatorname{Id}, \psi_{\dot{z}} \rangle_{{}^{n^{-1}}\mathbf{V}^F \cap \mathbf{U}^F}$ is not zero (and is equal to 1) if and only if ${}^{n^{-1}}\mathbf{V} \cap \mathbf{U}$ contains no $\mathbf{U}_\alpha$ for $\alpha \in \Pi$. However, as w is I-reduced, we have

$${}^{n^{-1}}\mathbf{V} \cap \mathbf{U} = {}^{n^{-1} n_0^I}\mathbf{U} \cap \mathbf{U},$$

where n_0^I is a representative of the longest element w_0^I of W_I (as the positive roots mapped by w_0^I to negative roots are exactly the roots of $<I>$). This group contains no $\mathbf{U}_\alpha$ with $\alpha \in \Pi$ if and only if $(w_0^I)^{-1}w$ maps any positive simple root to a negative root which means that it is the longest element w_0 of W, so that $w = w_0^I w_0$. As w_0 maps all positive roots to negative roots and w_0^I changes the sign of exactly those roots which are in the root system of $\mathbf{L}$, we have $\mathbf{L} \cap {}^n\mathbf{U} = \mathbf{L} \cap \mathbf{U}$, so

$$\langle R_{\mathbf{L}}^{\mathbf{G}}(f), \Gamma_z \rangle_{\mathbf{G}^F} = \langle \operatorname{Res}_{\mathbf{L}^F \cap \mathbf{U}^F}^{\mathbf{L}^F} f, {}^n\psi_{\dot z} \rangle_{\mathbf{L}^F \cap \mathbf{U}^F}.$$

So we shall have proved the proposition if we show that ${}^n\psi_{\dot z}|_{\mathbf{L}^F \cap \mathbf{U}^F}$ is in the $\mathbf{T}^F$-orbit $\Psi_{h_{\mathbf{L}}(z)}$ of regular characters of $\mathbf{L}^F \cap \mathbf{U}^F$. But we have $\psi_{\dot z} = {}^t\psi_1$ for any $t \in \mathcal{L}_{\mathbf{T}}^{-1}(Z(\mathbf{G}))$ such that $\dot z = tZ(\mathbf{G})$ and, by the definition of $h_{\mathbf{L}}$, the character ${}^t(\operatorname{Res}_{\mathbf{U}^F \cap \mathbf{L}^F}^{\mathbf{U}^F} \psi_1)$ is in $\Psi_{h_{\mathbf{L}}(z)}$. So, as $\mathcal{L}(nt) = \mathcal{L}(t)$, it is sufficient to show the result for $z = 1$. But then, by the choice of ψ_1, the result is clear. ■

We now give two results on the dual of a Gelfand-Graev representation.

14.33 PROPOSITION.

(i) *The dual of any Gelfand-Graev representation is zero outside regular unipotent elements.*

(ii) *We have* $\langle \Gamma_z, (-1)^{|\Pi/\tau|} D_{\mathbf{G}} \Gamma_{z'} \rangle_{\mathbf{G}^F} = |Z(\mathbf{G})^F| \delta_{z,z'}$.

PROOF: By 14.32 we have

$$D_{\mathbf{G}}\Gamma_z = \sum_{J \subset \Pi/\tau} (-1)^{|J|} R_{\mathbf{L}_J}^{\mathbf{G}}(\Gamma_{h_{\mathbf{L}_J}(z)}).$$

By the definition of $R_{\mathbf{L}}^{\mathbf{G}}$ we have

$$R_{\mathbf{L}_J}^{\mathbf{G}}(\Gamma_{h_{\mathbf{L}_J}(z)}) = \operatorname{Ind}_{\mathbf{P}_J^F}^{\mathbf{G}^F}[\operatorname{Ind}_{\mathbf{U}^F \cap \mathbf{L}_J^F}^{\mathbf{L}_J^F}(\operatorname{Res}_{\mathbf{U}^F \cap \mathbf{L}_J^F}^{\mathbf{U}^F}(\psi_{\dot z})) \times \operatorname{Id}_{\mathbf{V}_J^F}],$$

(where $\mathbf{V}_J$ is the unipotent radical of $\mathbf{P}_J$). But an easy computation shows that

$$\operatorname{Ind}_{\mathbf{U}^F \cap \mathbf{L}_J^F}^{\mathbf{L}_J^F}(\operatorname{Res}_{\mathbf{U}^F \cap \mathbf{L}_J^F}^{\mathbf{U}^F}(\psi_{\dot z})) \times \operatorname{Id}_{\mathbf{V}_J^F} = \operatorname{Ind}_{\mathbf{U}^F}^{\mathbf{P}_J^F}(\operatorname{Res}_{\mathbf{U}^F \cap \mathbf{L}_J^F}^{\mathbf{U}^F}(\psi_{\dot z}) \times \operatorname{Id}_{\mathbf{V}_J^F}),$$

so we get by transitivity of induction $R_{\mathbf{L}_J}^{\mathbf{G}}(\Gamma_{h_{\mathbf{L}_J}(z)}) = \operatorname{Ind}_{\mathbf{U}^F}^{\mathbf{G}^F}(\operatorname{Res}_{\mathbf{U}^F \cap \mathbf{L}_J^F}^{\mathbf{U}^F}(\psi_{\dot z}) \times \operatorname{Id}_{\mathbf{V}_J^F})$. Thus

$$D_{\mathbf{G}}(\Gamma_z) = \operatorname{Ind}_{\mathbf{U}^F}^{\mathbf{G}^F}\Big(\sum_{J \subset \Pi/\tau} (-1)^{|J|} \operatorname{Res}_{\mathbf{U}^F \cap \mathbf{L}_J^F}^{\mathbf{U}^F} \psi_{\dot z} \times \operatorname{Id}_{\mathbf{V}_J^F}\Big).$$

To prove assertion (i) of the proposition it is sufficient to show that

$$\sum_{J\subset\Pi/\tau}(-1)^{|J|}\operatorname{Res}^{\mathbf{U}^F}_{\mathbf{U}^F\cap\mathbf{L}_J^F}\psi_{\dot z}\times\operatorname{Id}_{\mathbf{V}_J^F}$$

is zero outside regular unipotent elements of $\mathbf{U}^F$. Let u be in $\mathbf{U}^F$ and let $\prod_{\mathcal{O}\in\Pi/\tau}u_{\mathcal{O}}$ denote the projection of u on $\prod_{\mathcal{O}}\mathbf{U}_{\mathcal{O}}^F$; if we put $\psi_{\mathcal{O}}=\operatorname{Res}^{\mathbf{U}^F}_{\mathbf{U}^F_{\mathcal{O}}}\psi_{\dot z}$, we have

$$\psi_{\dot z}(u)=\prod_{\mathcal{O}\in\Pi/\tau}\psi_{\mathcal{O}}(u_{\mathcal{O}})$$

whence

$$\sum_{J\subset\Pi/\tau}(-1)^{|J|}(\operatorname{Res}^{\mathbf{U}^F}_{\mathbf{U}^F\cap\mathbf{L}_J^F}\psi_{\dot z}\times\operatorname{Id}_{\mathbf{V}_J^F})(u)=\prod_{\mathcal{O}\in\Pi/\tau}(1-\psi_{\mathcal{O}}(u_{\mathcal{O}}))$$

which is zero outside regular unipotent elements, as u is regular if and only if $u_{\mathcal{O}}\neq 1$ for all $\mathcal{O}$.

We now prove (ii). By the above computations we have $D_{\mathbf{G}}\Gamma_z=\operatorname{Ind}^{\mathbf{G}^F}_{\mathbf{U}^F}\varphi$, where

$$\varphi=\sum_{J\subset\Pi/\tau}(-1)^{|J|}\operatorname{Res}^{\mathbf{U}^F}_{\mathbf{U}^F\cap\mathbf{L}_J^F}\psi_{\dot z}\times\operatorname{Id}_{\mathbf{V}_J^F},$$

which is zero outside regular unipotent elements. By the Mackey formula, using the fact that $N_{\mathbf{G}}(\mathbf{T})^F$ is a cross section for $\mathbf{U}^F\backslash\mathbf{G}^F/\mathbf{U}^F$, we have

$$\langle D_{\mathbf{G}}\Gamma_z,\Gamma_{z'}\rangle_{\mathbf{G}^F}=\sum_{n\in N_{\mathbf{G}}(\mathbf{T})^F}\langle\varphi,{}^n\psi_{\dot z'}\rangle_{\mathbf{U}^F\cap{}^n\mathbf{U}^F}.$$

But $\mathbf{U}^F\cap{}^n\mathbf{U}^F$ does not contain any regular unipotent element if n is not in $\mathbf{T}$, and $\langle\operatorname{Res}^{\mathbf{U}^F}_{\mathbf{U}^F\cap\mathbf{L}_J^F}\psi_{\dot z}\times\operatorname{Id}_{\mathbf{V}_J^F},\psi_{\dot z'}\rangle_{\mathbf{U}^F}$ is equal to zero if $J\neq\Pi/\tau$. So we have

$$\langle D_{\mathbf{G}}\Gamma_z,\Gamma_{z'}\rangle_{\mathbf{G}^F}=(-1)^{|\Pi/\tau|}\sum_{t\in\mathbf{T}^F}\langle\psi_{\dot z},{}^t\psi_{\dot z'}\rangle_{\mathbf{U}^F}.$$

The characters $\psi_{\dot z}$ and ${}^t\psi_{\dot z'}$ can be equal only if $z=z'$. There are then $|Z(\mathbf{G})^F|$ values of t such that $\psi_{\dot z}={}^t\psi_{\dot z'}$, whence the result. ∎

Using the Gelfand-Graev representations, we can give the value of any character on a regular unipotent element at least when the characteristic is good.

14.34 DEFINITION. *We put* $\sigma_z=\sum_{\psi\in\Psi_{z^{-1}}}\psi(u_1)$.

14.35 THEOREM.
(i) For any $\chi\in\operatorname{Irr}(\mathbf{G}^F)$ *we have*

$$|\mathcal{U}_z|^{-1}\sum_{u\in\mathcal{U}_z}\chi(u)=\sum_{z'\in H^1(F,\mathbf{Z})}\sigma_{zz'^{-1}}\langle(-1)^{|\Pi/\tau|}D_{\mathbf{G}}(\chi),\Gamma_{z'}\rangle_{\mathbf{G}^F}.$$

(ii) *We have* $\langle \sum_z \Gamma_z, \sum_z \Gamma_z \rangle = |H^1(F, Z(\mathbf{G}))||Z(\mathbf{G})^F|q^l$, *where l is the semi-simple rank of* $\mathbf{G}$.

PROOF: In the following we shall denote by γ_z the class function on $\mathbf{G}^F$ whose value is zero outside $\mathcal{U}_z$ and $|\mathbf{G}^F|/|\mathcal{U}_z|$ on $\mathcal{U}_z$. We have seen in the proof of 14.33 that $D_{\mathbf{G}}\Gamma_z = \mathrm{Ind}_{\mathbf{U}^F}^{\mathbf{G}^F} \varphi$, where φ is a function on $\mathbf{U}^F/\mathbf{U}^{*F}$ which is zero outside the set of regular elements. So by 14.26 $D_{\mathbf{G}}\Gamma_z$ is constant on $\mathcal{U}_{z'}$, and there exist coefficients $c_{z,z'}$ such that $D_{\mathbf{G}}\Gamma_z = \sum_{z' \in H^1(F,\mathbf{Z})} c_{z,z'}\gamma_{z'}$. By 14.33 (ii) the matrix $(c_{z,z'})_{z,z'}$ is invertible and its inverse is

$$((-1)^{|\Pi/\tau|}|Z(\mathbf{G})^F|^{-1}\langle \Gamma_{z'}, \gamma_z \rangle_{\mathbf{G}^F})_{z,z'}.$$

Let $t \in \mathcal{L}_{\mathbf{T}}^{-1}(\mathbf{Z})$ be such that the F-class of $\mathcal{L}(t)$ is equal to z; we have

$$\langle \Gamma_{z'}, \gamma_z \rangle_{\mathbf{G}^F} = |\mathcal{U}_z|^{-1} \sum_{u \in \mathcal{U}_z} \Gamma_{z'}(u) = \Gamma_{z'}({}^t u_1),$$

the last equality because $\Gamma_{z'}$ is constant on $\mathcal{U}_z$, as it is induced from $\psi_{\dot{z}'}$ which factorizes through $(\mathbf{U}/\mathbf{U}^*)^F$. So

$$\langle \Gamma_{z'}, \gamma_z \rangle_{\mathbf{G}^F} = \Gamma_{z'}({}^t u_1) = \Gamma_{z'z^{-1}}(u_1).$$

We now apply the following lemma:

14.36 LEMMA. *We have* $\sigma_z = |Z(\mathbf{G})^F|^{-1}\Gamma_{z^{-1}}(u_1)$.

PROOF: By definition

$$\Gamma_z(u_1) = |\mathbf{U}^F|^{-1} \sum_{\{g \in \mathbf{G}^F \mid {}^g u_1 \in \mathbf{U}^F\}} \psi_{\dot{z}}({}^g u_1).$$

As u_1 is regular, we must have $g \in \mathbf{B}^F$ in the above summation and, if $g = tv$ with $t \in \mathbf{T}^F$ and $v \in \mathbf{U}^F$, then $\psi_{\dot{z}}({}^g u_1) = \psi_{\dot{z}}({}^t u_1)$, whence, as $Z(\mathbf{G})^F$ is the kernel of the $\mathbf{T}^F$-action on Ψ_z, we get

$$|Z(\mathbf{G})^F|^{-1}\Gamma_z(u_1) = \sum_{\psi \in \Psi_z} \psi(u_1) = \sigma_{z^{-1}}.$$

■

So we get

$$|Z(\mathbf{G})^F|^{-1}\langle \Gamma_{z'}, \gamma_z \rangle_{\mathbf{G}^F} = \sigma_{zz'^{-1}}.$$

So $(-1)^{|\Pi/\tau|}(\sigma_{zz'^{-1}})_{z,z'}$ is the inverse matrix of $(c_{z,z'})_{z,z'}$, *i.e.*,

$$\gamma_z = \sum_{z' \in H^1(F,\mathbf{Z})} \sigma_{zz'^{-1}}(-1)^{|\Pi/\tau|} D_{\mathbf{G}}\Gamma_{z'},$$

whence (i) by taking the scalar product of both sides with χ.

Let us prove (ii). We have $\langle \sum_z \Gamma_z, \sum_z \Gamma_z \rangle = \langle \sum_z D_{\mathbf{G}}\Gamma_z, \sum_z D_{\mathbf{G}}\Gamma_z \rangle$ and, with the above notation, it follows that

$$\begin{aligned}\langle \sum_z D_{\mathbf{G}}\Gamma_z, \sum_z D_{\mathbf{G}}\Gamma_z \rangle &= \langle \sum_{z,z'} c_{z,z'}\gamma_{z'}, \sum_{z,z'} c_{z,z'}\gamma_{z'} \rangle \\ &= \sum_{z'} \langle \sum_z c_{z,z'}\gamma_{z'}, \sum_z c_{z,z'}\gamma_{z'} \rangle.\end{aligned}$$

But

$$\begin{aligned}\sum_{z'} c_{z,z'} &= \langle D_{\mathbf{G}}\Gamma_z, \mathrm{Id}_{\mathbf{G}} \rangle = \langle \Gamma_z, \mathrm{St}_{\mathbf{G}} \rangle = |\mathbf{G}^F|^{-1}\Gamma_z(1)\,\mathrm{St}_{\mathbf{G}}(1) \\ &= |\mathbf{G}^F|^{-1}(|\mathbf{G}^F|/|\mathbf{U}^F|)|\mathbf{G}^F|_p = 1 \text{ (see 3.19 (i))},\end{aligned}$$

whence $\sum_z c_{z,z'} = 1$ as $c_{z,z'} = c_{1,z^{-1}z'}$. Thus

$$\langle \sum_z D_{\mathbf{G}}\Gamma_z, \sum_z D_{\mathbf{G}}\Gamma_z \rangle = \sum_{z'} \langle \gamma_{z'}, \gamma_{z'} \rangle = \sum_{z'} |\mathbf{G}^F|/|\mathcal{U}_{z'}|.$$

All the sets $\mathcal{U}_z$ have the same cardinality as they are geometrically conjugate, so their cardinality is $|H^1(F, Z(\mathbf{G}))|^{-1}$ times the number of regular unipotent elements. Hence the above sum is equal to $|Z(\mathbf{G})^{\circ F}|q^l|H^1(F, Z(\mathbf{G}))|^2$ by 14.23. As $Z(\mathbf{G})/Z(\mathbf{G})^\circ$ is abelian we have

$$|H^1(F, Z(\mathbf{G}))| = |H^1(F, Z(\mathbf{G})/Z(\mathbf{G})^\circ)| = |Z(\mathbf{G})^F/Z(\mathbf{G})^{\circ F}|,$$

whence the result. ■

14.37 COROLLARY. *If the centre of* $\mathbf{G}$ *is connected and* Γ *denotes the unique Gelfand-Graev representation, then* $D_{\mathbf{G}}\Gamma$ *is equal to* $|Z(\mathbf{G})^F|q^l$ *on regular unipotent elements (and to zero elsewhere). If moreover the characteristic is good we have* $D_{\mathbf{G}}\Gamma = \pi_u^{\mathbf{G}^F}$, *where* u *is a rational regular unipotent element.*

PROOF: In that case there is only one $c_{z,z'}$ whose value has to be 1 by the proof of 14.35 (ii). If moreover the characteristic is good the set of rational regular unipotent elements is one conjugacy class. ■

14.38 COROLLARY. *If the characteristic is good for* $\mathbf{G}$, *we have*

$$\chi(u) = \sum_{z' \in H^1(F,\mathbf{Z})} \sigma_{zz'^{-1}} \langle (-1)^{|\Pi/\tau|} D_{\mathbf{G}}(\chi), \Gamma_{z'} \rangle_{\mathbf{G}^F}$$

for any $u \in \mathcal{U}_z$.

PROOF: In that case the set $\mathcal{U}_z$ is one conjugacy class by 14.18. ■

Note that, since for bad characteristic the distinct conjugacy classes into which $\mathcal{U}_z$ splits have all the same cardinality (see 14.22), the left-hand side of 14.35 is the mean of the values of χ at these classes.

The final part of this chapter is devoted to the study of irreducible components of Gelfand-Graev representations and of their dual. We shall prove in particular that when the centre of $\mathbf{G}$ is connected these representations are uniform.

14.39 DEFINITION. *An irreducible character of* $\mathbf{G}^F$ *is* **regular** *if it is a component of some Gelfand-Graev character. An irreducible character whose dual is (up to sign) a regular irreducible character will be called* **semi-simple**.

For example the character $\mathrm{St}_{\mathbf{G}}$ is regular as $\mathrm{Res}_{\mathbf{U}^F}^{\mathbf{G}^F} \mathrm{St}_{\mathbf{G}} = \mathrm{reg}_{\mathbf{U}^F}$ (see 9.3 and 3.19 (i)), so by Frobenius reciprocity $\mathrm{St}_{\mathbf{G}}$ is a component of any Gelfand-Graev representation. The character $\mathrm{Id}_{\mathbf{G}}$ is thus semi-simple. The following definition will allow us to describe the regular and semi-simple characters (see 14.47 and 14.49 below).

14.40 DEFINITION. *If s is a semi-simple element of $\mathbf{G}^{*F^*}$ and if $\mathbf{T}^*$ is a rational maximal torus containing s, we define a class-function $\chi_{(s)}$ on $\mathbf{G}^F$ by*

$$\chi_{(s)} = |W^\circ(s)|^{-1} \sum_{w \in W^\circ(s)} \varepsilon_{\mathbf{G}} \varepsilon_{\mathbf{T}^*_w} R^{\mathbf{G}}_{\mathbf{T}^*_w}(s),$$

where $W^\circ(s)$ is as in 2.4 and $\mathbf{T}^*_w$ is a torus of $\mathbf{G}^*$ of type w (with respect to $\mathbf{T}^*$; see notation after 13.13).

Note that $\varepsilon_{\mathbf{G}} \varepsilon_{\mathbf{T}^*_w} = \varepsilon_{\mathbf{T}^*}(-1)^{l(w)}$. Note also that the above definition does not depend on $\mathbf{T}^*$: choosing another rational maximal torus containing s only makes a translation on the types inside $W^\circ(s)$.

We now define a **rational series** of characters, denoted by $\mathcal{E}(\mathbf{G}^F, (s)_{\mathbf{G}^{*F^*}})$, as the set of irreducible components of the $R^{\mathbf{G}}_{\mathbf{T}^*}(s)$ where the semi-simple class $(s)_{\mathbf{G}^{*F^*}}$ of s in $\mathbf{G}^{*F^*}$ is fixed; it is thus a subset of $\mathcal{E}(\mathbf{G}^F, (s))$. Then we see that $\chi_{(s)}$ is in $\mathcal{E}(\mathbf{G}^F, (s)_{\mathbf{G}^{*F^*}})$. Note that by 13.15 (ii) and 3.21, if the centre of $\mathbf{G}$ is connected, the rational series and the geometric series are the same. This is not true in general, as can be seen from the following proposition.

14.41 PROPOSITION. *The rational series of characters form a partition of* $\mathrm{Irr}(\mathbf{G}^F)$.

This result will be proved later (see 14.50). We shall also use the next result.

14.42 PROPOSITION. *The number of rational geometric semi-simple classes of* $\mathbf{G}^*$ *is* $|Z(\mathbf{G})^{\circ F}|q^l$, *where* l *is the semi-simple rank of* $\mathbf{G}$.

SKETCH OF THE PROOF: We follow the proof given in [DL1, 5.7]. Let $\mathbf{T}^*$ be the rational maximal torus of $\mathbf{G}^*$ which defines the duality with $\mathbf{G}$. Any semi-simple class has a representative in $\mathbf{T}^*$ and two elements of $\mathbf{T}^*$ are conjugate in $\mathbf{G}^*$ if and only if they are conjugate by the Weyl group W of $\mathbf{G}^*$ with respect to $\mathbf{T}^*$ (see 0.12 (iv)). So geometric semi-simple conjugacy classes are in one-to-one correspondence with $\mathbf{T}^*/W$ and rational semi-simple geometric classes are in one-to-one correspondence with $(\mathbf{T}^*/W)^{F^*}$. By 10.4 and 10.10 (i) we have

$$|(\mathbf{T}^*/W)^{F^*}| = \mathrm{Trace}(F^*|H_c^*(\mathbf{T}^*/W)) = \mathrm{Trace}(F^*|H_c^*(\mathbf{T}^*)^W).$$

We then apply Künneth's theorem 10.9 to the decomposition $\mathbf{T}^* = \mathbf{T}^{*\prime}.Z(\mathbf{G}^*)^\circ$ where $\mathbf{T}^{*\prime}$ is the intersection of $\mathbf{T}^*$ with the derived group of $\mathbf{G}^*$. It is then easy to see that the trace is equal to $|Z(\mathbf{G}^*)^{\circ F^*}|\,\mathrm{Trace}(F^*|H_c^*(\mathbf{T}^{*\prime})^W)$. We have $|Z(\mathbf{G}^*)^{\circ F^*}| = |Z(\mathbf{G})^{\circ F}|$. Using Poincaré duality and the fact that $\mathbf{T}^{*\prime} \simeq Y(\mathbf{T}^{*\prime}) \otimes (\mathbb{Q}/\mathbb{Z})_{p'}$ one gets

$$\mathrm{Trace}(F^*|H_c^{2l-i}(\mathbf{T}^{*\prime})^W) = q^{l-i}\,\mathrm{Trace}(F, \wedge^i Y(\mathbf{T}^{*\prime})^W).$$

A known result on the action of a reflection group on a real vector space (see [Bbk V, ex. 3 of §2]) shows that W has no fixed point on $\wedge^i Y(\mathbf{T}^{*\prime})$ if $i \neq 0$, which gives the result. ∎

14.43 PROPOSITION. *We have* $\langle \chi_{(s)}, \chi_{(s)} \rangle_{\mathbf{G}^F} = |(W(s)/W^\circ(s))^{F^*}|$.

PROOF: By definition

$$\begin{aligned}\langle \chi_{(s)}, \chi_{(s)} \rangle &= |W^\circ(s)|^{-2} \sum_{w,w' \in W^\circ(s)} \langle R^{\mathbf{G}}_{\mathbf{T}^*_w}(s), R^{\mathbf{G}}_{\mathbf{T}^*_{w'}}(s) \rangle \\ &= |W^\circ(s)|^{-2} \sum_{w \in W^\circ(s)} |W(s)^{wF^*}|\,|W^\circ(s) \cap \{F^*\text{-class of } w \text{ in } W(s)\}| \\ &= |W^\circ(s)|^{-2} \sum_{w \in W^\circ(s)} |\{ v \in W(s) \mid vw^{F^*}v^{-1} \in W^\circ(s) \}|,\end{aligned}$$

the second equality by 11.15 (as the proof of 13.13 shows that the stabilizer of θ in $W(\mathbf{T})^F$ is isomorphic to that of s in $W(\mathbf{T}^*)^{F^*}$ for corresponding pairs $(\mathbf{T}, \theta)$ and $(\mathbf{T}^*, s)$). But, as w is in $W^\circ(s)$, the F^*-conjugate by $v \in W(s)$ of w is in $W^\circ(s)$ if and only if the image of v in $W(s)/W^\circ(s)$ is F^*-fixed, so we get

$$\langle \chi_{(s)}, \chi_{(s)} \rangle = |W^\circ(s)|^{-2} \sum_{w \in W^\circ(s)} |(W(s)/W^\circ(s))^{F^*}|\,|W^\circ(s)|,$$

whence the result. ■

14.44 PROPOSITION. *For any $z \in H^1(F, Z(\mathbf{G}))$ we have $\langle \chi_{(s)}, \Gamma_z \rangle_{\mathbf{G}^F} = 1$.*

PROOF: We shall use the following property of Green functions.

14.45 LEMMA. *The value of any Deligne-Lusztig character at a regular unipotent element is 1.*

PROOF: The proof is in [DL1, 9.16]. Its main ingredient is the use of a compactification of the variety $\mathcal{L}^{-1}(\mathbf{U})/\mathbf{T}$. The trace of u on $H_c^*(\mathcal{L}^{-1}(\mathbf{U})/\mathbf{T})$ is then computed, using the fact that u, being regular unipotent, has only one fixed point a on that compactification, and by taking local coordinates at a. ■

By definition we have

$$\langle \chi_{(s)}, \Gamma_z \rangle = |W^\circ(s)|^{-1} \sum_{w \in W^\circ(s)} \langle \varepsilon_{\mathbf{G}} \varepsilon_{\mathbf{T}_w^*} R_{\mathbf{T}_w^*}^{\mathbf{G}}(s), \Gamma_z \rangle.$$

But

$$\varepsilon_{\mathbf{G}} \varepsilon_{\mathbf{T}_w^*} R_{\mathbf{T}_w^*}^{\mathbf{G}}(s) = D_{\mathbf{G}}(R_{\mathbf{T}_w^*}^{\mathbf{G}}(s))$$

by 12.8 and the proof of 13.13, so the above scalar product is equal to

$$\langle D_{\mathbf{G}}(R_{\mathbf{T}_w^*}^{\mathbf{G}}(s)), \Gamma_z \rangle = \langle R_{\mathbf{T}_w^*}^{\mathbf{G}}(s), D_{\mathbf{G}}(\Gamma_z) \rangle.$$

If we express $D_{\mathbf{G}}(\Gamma_z)$ as linear combinations of the $\gamma_{z'}$ as in the proof of 14.35, we get $\sum_{z'} c_{z,z'} \langle R_{\mathbf{T}_w^*}^{\mathbf{G}}(s), \gamma_{z'} \rangle$; but, by 14.45, $\langle R_{\mathbf{T}_w^*}^{\mathbf{G}}(s), \gamma_{z'} \rangle$ is equal to 1 and by the proof or 14.35, $\sum_{z'} c_{z,z'}$ is equal to 1, whence the result. ■

We can now prove

14.46 PROPOSITION. *We have*

$$|Z(\mathbf{G})^F| / |Z(\mathbf{G})^{\circ F}| \sum_{(s)} |(W(s)/W^\circ(s))^{F^*}|^{-1} \chi_{(s)} = \sum_{z \in H^1(F, Z(\mathbf{G}))} \Gamma_z,$$

where in the sum of the left-hand side (s) runs over rational semi-simple conjugacy classes of $\mathbf{G}^{*F^*}$.

PROOF: To prove the result we shall show that the scalar product of the two sides is equal to the scalar product of each side with itself. We first compute the scalar product of the left-hand side with itself. By 14.41 this is equal to

$$|Z(\mathbf{G})^F|^2 / |Z(\mathbf{G})^{\circ F}|^2 \sum_{(s)} |(W(s)/W^\circ(s))^{F^*}|^{-2} \langle \chi_{(s)}, \chi_{(s)} \rangle,$$

which, by 14.43 equals

$$|Z(\mathbf{G})^F|^2/|Z(\mathbf{G})^{\circ F}|^2 \sum_{(s)} |(W(s)/W^\circ(s))^{F^*}|^{-2} |(W(s)/W^\circ(s))^{F^*}| = \\ |Z(\mathbf{G})^F|^2/|Z(\mathbf{G})^{\circ F}|^2 \sum_{(s)} |(W(s)/W^\circ(s))^{F^*}|^{-1}.$$

As $|(W(s)/W^\circ(s))^{F^*}|$ is the number of rational classes in the geometric conjugacy class of s, the last sum is equal to

$$|Z(\mathbf{G})^F|^2/|Z(\mathbf{G})^{\circ F}|^2 \#\{\text{rational geometric semi-simple classes of } \mathbf{G}^*\}.$$

By 14.42, this equals

$$|Z(\mathbf{G})^F||Z(\mathbf{G})^F/Z(\mathbf{G})^{\circ F}|q^l,$$

which can be written

$$|Z(\mathbf{G})^F||H^1(F, Z(\mathbf{G}))|q^l.$$

The scalar product of the right-hand side with itself was computed in 14.35 and has the same value. The scalar product of one side with the other one is easily computed from 14.44: it is the product of $|Z(\mathbf{G})^F/Z(\mathbf{G})^{\circ F}|$, the number of rational geometric conjugacy classes and the number of z, which gives again $|Z(\mathbf{G})^F|q^l|H^1(F, Z(\mathbf{G}))|$. ■

14.47 COROLLARY. *Assume that the centre of* $\mathbf{G}$ *is connected. Then:*

(i) $\chi_{(s)}$ *is an irreducible character of* $\mathbf{G}^F$ *for any* (s),

(ii) The (unique) Gelfand-Graev representation of $\mathbf{G}^F$ *is the sum of all the* $\chi_{(s)}$ *(which are all the regular characters),*

(iii) The dual of the Gelfand-Graev representation (which is $\pi_u^{\mathbf{G}^F}$ *if the characteristic is good; see 14.37) is the sum of all the semi-simple characters of* $\mathbf{G}^F$ *up to signs.*

PROOF: By hypothesis, from 13.15 (ii), all the numbers $|(W(s)/W^\circ(s))^{F^*}|$ are equal to 1 and $\langle \chi_{(s)}, \chi_{(s)} \rangle = 1$ (see 14.43). The corollary is then straightforward (note that in the present case the disjointness of the rational series 14.41 reduces to the disjointness of the geometric series, so that we do not actually have to use 14.41 here). ■

We now want to get a result similar to 14.47 for groups with a non-connected centre. First we shall imbed $\mathbf{G}$ in a group with connected centre and the same derived group. The construction is as follows: take any torus $\mathbf{S}$ containing $Z(\mathbf{G})$ and put $\tilde{\mathbf{G}} = \mathbf{G} \times_{\mathbf{Z}} \mathbf{S}$, where $\mathbf{Z}$ acts by translation on $\mathbf{G}$ and on

$\mathbf{S}$. The centre of $\tilde{\mathbf{G}}$ is isomorphic to $\mathbf{S}$ and the derived group of $\tilde{\mathbf{G}}$ is clearly the same as that of $\mathbf{G}$. If $\mathbf{T}$ is a maximal torus of $\mathbf{G}$ then it is contained in a unique maximal torus $\tilde{\mathbf{T}} = \mathbf{T} \times_{\mathbf{Z}} \mathbf{S}$ (see 0.6) of $\tilde{\mathbf{G}}$ and $\mathbf{T} = \tilde{\mathbf{T}} \cap \mathbf{G}$. We shall assume in addition that $\mathbf{S}$ is rational, and extend F to $\tilde{\mathbf{G}}$ in the obvious way. We then have a bijection from the set of rational maximal tori of $\mathbf{G}$ onto that of $\tilde{\mathbf{G}}$. If θ is a character of $\tilde{\mathbf{T}}^F$ (where $\mathbf{T}$ is a rational maximal torus of $\mathbf{G}$) then by 13.22 the restriction of $R_{\tilde{\mathbf{T}}}^{\tilde{\mathbf{G}}}(\theta)$ to $\mathbf{G}^F$ is $R_{\mathbf{T}}^{\mathbf{G}}(\mathrm{Res}_{\mathbf{T}^F}^{\tilde{\mathbf{T}}^F} \theta)$.

We now consider dual groups. If $\tilde{\mathbf{T}}^*$ is a torus dual to $\tilde{\mathbf{T}}$ and $\mathbf{T}^*$ a torus dual to $\mathbf{T}$, then $X(\mathbf{T})$ is a quotient of $X(\tilde{\mathbf{T}})$, so $Y(\mathbf{T}^*)$ is a quotient of $Y(\tilde{\mathbf{T}}^*)$, so $\mathbf{T}^*$ is naturally a quotient of $\tilde{\mathbf{T}}^*$ by 0.20. Let $\tilde{\mathbf{G}}^*$ be a group dual to $\tilde{\mathbf{G}}$ (the torus $\tilde{\mathbf{T}}^*$ of $\tilde{\mathbf{G}}^*$ being dual to the torus $\tilde{\mathbf{T}}$ of $\tilde{\mathbf{G}}$); then the kernel of $\tilde{\mathbf{T}}^* \to \mathbf{T}^*$ is a central torus of $\tilde{\mathbf{G}}^*$ and the quotient $\mathbf{G}^*$ of $\tilde{\mathbf{G}}^*$ by this central torus is clearly dual to $\mathbf{G}$ (the torus $\mathbf{T}^*$ being dual to $\mathbf{T}$). We shall assume also that $\tilde{\mathbf{G}}^*$ has a Frobenius endomorphism F^* dual to F, so the same is true for $\mathbf{G}^*$. Note that the root systems of $\tilde{\mathbf{G}}^*$ (relative to $\tilde{\mathbf{T}}^*$) and of $\mathbf{G}^*$ (relative to $\mathbf{T}^*$) are the same.

14.48 PROPOSITION. *For any $s \in \mathbf{G}^{*F^*}$ the class function $\chi_{(s)}$ on $\mathbf{G}^F$ is a proper character.*

PROOF: Let $\tilde{s} \in \tilde{\mathbf{G}}^{*F^*}$ be a semi-simple element whose image is $s \in \mathbf{G}^{*F^*}$ ($\tilde{s}$ exists as the kernel of $\tilde{\mathbf{G}}^* \to \mathbf{G}^*$ is connected); we shall show that $\chi_{(s)}$ is the restriction of the character $\chi_{(\tilde{s})}$ of $\tilde{\mathbf{G}}^F$, which implies the result by 14.47 (i). The group $W(\tilde{s})$ is the group generated by the roots of $\tilde{\mathbf{G}}^*$ which are trivial on $\tilde{s}$ (as the centre of $\tilde{\mathbf{G}}$ is connected, hence centralizers of semi-simple elements in $\tilde{\mathbf{G}}^*$ are connected); so this group is the same as $W^\circ(s)$. Since $\tilde{s}$ maps to s, the character $R_{\mathbf{T}^*}^{\mathbf{G}}(s)$ is the restriction to $\mathbf{G}^F$ of the character $R_{\tilde{\mathbf{T}}}^{\tilde{\mathbf{G}}}(\tilde{s})$, and, from the definitions of $\chi_{(s)}$ and $\chi_{(\tilde{s})}$ the result follows. ■

Once we know that $\chi_{(s)}$ is a proper character, we can give the decomposition of the Gelfand-Graev representation in general.

14.49 THEOREM. *For any $z \in H^1(F, Z(\mathbf{G}))$ and any rational semi-simple conjugacy class of $\mathbf{G}^{*F^*}$, there is exactly one irreducible common component $\chi_{s,z}$ of $\chi_{(s)}$ and Γ_z; it has multiplicity 1 in both $\chi_{(s)}$ and Γ_z; and we have*

$$\Gamma_z = \sum_{(s)} \chi_{s,z},$$

where (s) runs over the set of semi-simple classes of $\mathbf{G}^{*F^*}$.

PROOF: The result is straightforward by 14.44 and 14.46. ■

Note that we have not given a parametrization of the regular characters: indeed it is possible to have $\chi_{s,z} = \chi_{s,z'}$. For instance, when $s = 1$ we have $\chi_{(s)} = \mathrm{St}_{\mathbf{G}}$ as observed in the proof of 12.14. It can be shown that $\chi_{s,z} = \chi_{s,z'}$ if and only if $z^{-1}z'$ is in a subgroup of $H^1(F, Z(\mathbf{G}))$ associated to s.

We now give a sketchy proof of 14.41.

14.50 PROOF OF 14.41: Let s be an element of $\mathbf{G}^{*F^*}$ and $\tilde{s}$ be a preimage of s in $\tilde{\mathbf{G}}^{*F^*}$. The series $\mathcal{E}(\mathbf{G}^F, (s)_{\mathbf{G}^{*F^*}})$ consists, by 13.22, of the irreducible components of the restrictions of elements in $\mathcal{E}(\tilde{\mathbf{G}}^F, (\tilde{s}))$. Assume that two series $\mathcal{E}(\mathbf{G}^F, (s)_{\mathbf{G}^{*F^*}})$ and $\mathcal{E}(\mathbf{G}^F, (s')_{\mathbf{G}^{*F^*}})$ have a non-empty intersection. Let $\tilde{s}'$ be a preimage in $\tilde{\mathbf{G}}^{*F^*}$ of s'; then there exist two irreducible characters $\chi \in \mathcal{E}(\tilde{\mathbf{G}}^F, (\tilde{s}))$ and $\chi' \in \mathcal{E}(\mathbf{G}^{*F^*}, (\tilde{s}'))$ whose restrictions to $\mathbf{G}^F$ have a common component. By Clifford's theory (see 13.21) we see that the restrictions of two irreducible characters χ and χ' of $\tilde{\mathbf{G}}^F$ to $\mathbf{G}^F$ are disjoint or equal and that they are equal if and only if $\chi' = \chi.\zeta$, where ζ is a linear character of $\tilde{\mathbf{G}}^F/\mathbf{G}^F$. We know that a character of $\tilde{\mathbf{G}}^F/\mathbf{G}^F$ is of the form $\hat{z}$ with $z \in \tilde{\mathbf{G}}^{*F^*}$ (see 13.30 as such a character is trivial on $\tilde{\mathbf{G}}'^F$). Since multiplication by $\hat{z}$ maps $R_{\mathbf{T}}^{\tilde{\mathbf{G}}}(\tilde{s})$ to $R_{\mathbf{T}}^{\tilde{\mathbf{G}}}(\tilde{s}z)$, the series $\mathcal{E}(\tilde{\mathbf{G}}^F, (\tilde{s}'))$ and $\mathcal{E}(\tilde{\mathbf{G}}^F, (\tilde{s}z))$ have to be equal as they have $\chi\hat{z}$ as a common component. But $\hat{z}$ is trivial on $\mathbf{G}^F$, so we see that the series $\mathcal{E}(\tilde{\mathbf{G}}^F, (\tilde{s}))$ and $\mathcal{E}(\tilde{\mathbf{G}}^F, (\tilde{s}z))$ have the same restrictions, and thus $\mathcal{E}(\mathbf{G}^F, (s)_{\mathbf{G}^{*F^*}}) = \mathcal{E}(\mathbf{G}^F, (s')_{\mathbf{G}^{*F^*}})$. ■

The following result gives the converse of 14.41.

14.51 THEOREM. *Let s and s' be two semi-simple elements of $\mathbf{G}^{*F^*}$; then two Deligne-Lusztig characters $R_{\mathbf{T}^*}^{\mathbf{G}}(s)$ and $R_{\mathbf{T}'^*}^{\mathbf{G}}(s')$ have a common component if and only if s and s' are rationally conjugate.*

PROOF: We have just seen the "only if" part. A similar computation to that of 14.43 shows that for any $w \in W^\circ(s)$ we have $\langle \chi_{(s)}, \chi_{(s)} \rangle = \langle \chi_{(s)}, R_{\mathbf{T}_w^*}^{\mathbf{G}}(s) \rangle$. As $R_{\mathbf{T}_w^*}^{\mathbf{G}}(s)$ is invariant by conjugation under $\mathcal{L}_{\mathbf{T}}^{-1}(Z(\mathbf{G}))$, all the irreducible components of $\chi_{(s)}$ have the same multiplicity in $R_{\mathbf{T}_w^*}^{\mathbf{G}}(s)$, so they must have multiplicity 1. Thus we see that, when s and s' are conjugate in $\mathbf{G}^{*F^*}$, all irreducible components of $\chi_{(s)}$ $(= \chi_{(s')})$ occur in both $R_{\mathbf{T}^*}^{\mathbf{G}}(s)$ and $R_{\mathbf{T}'^*}^{\mathbf{G}}(s')$. ■

References

Many results about regular elements can be found in [St2]. The Gelfand-Graev representation was first introduced by Gelfand and Graev [Construction of irreducible representations of simple algebraic groups over a finite field, *Doklady Akad. Nauk SSSR* **147** (1962), 529–532] who proved that it is multiplicity free in the case of $\mathbf{SL}_n$. The general result is due to Steinberg. All results about semi-simple characters and the decomposition of the Gelfand-Graev representation in the case of a group with connected centre are due to Deligne and Lusztig ([DL1]).

15. EXAMPLES

In this chapter we look at several examples of finite groups of Lie type; we elucidate their structure as rational points of reductive groups, and in some cases give their character table. We start with the linear groups.

15.1 The linear and unitary groups

The linear group $\mathbf{GL}_n(k)$ over some algebraically closed field k is defined by the k-algebra $k[T_{i,j}, \det(T_{i,j})^{-1}]$ endowed with the comultiplication $T_{i,j} \mapsto \sum_k T_{i,k} \otimes T_{k,j}$ (see 0.1 (ii)). As an open subset of an affine space, its variety is affine and connected. The group of points over k is the group of invertible $n \times n$-matrices with coefficients in k. The subgroup $\mathbf{T}$ of diagonal matrices is clearly a torus; this torus is maximal since (by well-known results on linear transformations) any set of commuting semi-simple matrices is "diagonalizable", *i.e.*, conjugate to a subgroup of $\mathbf{T}$. The subgroup $\mathbf{B}$ of upper triangular matrices is solvable, and is the semi-direct product $\mathbf{U} \rtimes \mathbf{T}$, where $\mathbf{U}$ is the nilpotent group of unipotent upper triangular matrices (*i.e.*, upper triangular matrices whose diagonal entries are all 1). The variety $\mathbf{U}$ is isomorphic to an affine space (of dimension $n(n-1)/2$), thus $\mathbf{B}$ is connected, and the Lie-Kolchin theorem (see 0.9) implies that it is a Borel subgroup. Similarly, the subgroup of lower triangular matrices is another Borel subgroup. As these two Borel subgroups intersect in a maximal torus $\mathbf{T}$, the group $\mathbf{GL}_n$ is reductive (see 0.32 (i)). However $\mathbf{GL}_n$ is not semi-simple, as its centre is the one-dimensional torus of all scalar matrices: it is connected and equal to the radical of $\mathbf{GL}_n$. The subgroup $\mathbf{U}_{i,j}$ of $\mathbf{U}$ of elements whose only non-zero off-diagonal coefficient is in position (i,j), where $i < j$, is clearly isomorphic to $\mathbf{G}_a$ and normalized by $\mathbf{T}$. It is thus a root subgroup $\mathbf{U}_\alpha$; the corresponding root α maps the element $\operatorname{diag}(t_1, \ldots, t_n)$ of $\mathbf{T}$ to t_i/t_j. The $n(n-1)/2$ positive roots, for the order defined by $\mathbf{B}$, thus obtained are clearly all the positive roots, as they are in number equal to $\dim(\mathbf{U})$. The simple roots for this order are $\Pi = \{\, \alpha_i \mid i = 1, \ldots, n-1\,\}$ where $\alpha_i(\operatorname{diag}(t_1 \ldots, t_n)) = t_i/t_{i+1}$. The root system thus obtained is of type A_{n-1}. The normalizer $N_{\mathbf{G}}(\mathbf{T})$ is the subgroup of monomial matrices, and the Weyl group is isomorphic to the symmetric group $\mathcal{S}_n$ (it acts on $\mathbf{T}$ by permuting the diagonal entries of a matrix). The parabolic subgroup $\mathbf{P}_J$

containing $\mathbf{B}$ where $J \subset \Pi$ is the subgroup of "block-triangular" matrices where the blocks correspond to maximal intervals $[i,k]$ in $[1,n]$ such that $\alpha_i,\ldots,\alpha_{k-1} \in J$.

When $k = \overline{\mathbb{F}}_q$, the group $\mathbf{GL}_n$ is endowed with the standard $\mathbb{F}_q$-structure where the Frobenius endomorphism F acts by raising all entries of a matrix to the q-th power (see 3.9). The group $\mathbf{GL}_n^F$ is then the linear group of $\mathbb{F}_q{}^n$. The group $\mathbf{GL}_n$ may also be endowed with the $\mathbb{F}_q$-structure corresponding to the Frobenius endomorphism F' given by ${}^{F'}g = {}^{tF}g^{-1}$ (where ${}^tg$ denotes the matrix which is the transpose of g). The group $\mathbf{GL}_n^{F'}$ is the group of unitary transformations of $(\mathbb{F}_{q^2})^n$. The group $\mathbf{GL}_n$ with the Frobenius endomorphism F' is called the unitary group and denoted by $\mathbf{U}_n$ (this notation is somewhat inconsistent as it is not a group but only a rational form of one). Let us compute the $\mathbb{F}_q$-rank of $\mathbf{U}_n$. The group ${}^{F'}\mathbf{B}$ is the Borel subgroup of lower triangular matrices, and is also equal to ${}^{w_0}\mathbf{B}$, where w_0 is the longest element of $W(\mathbf{T})$ (which is the permutation $(1,2,\ldots,n) \mapsto (n,n-1,\ldots,1)$). Thus $\mathbf{T}$ is of type w_0 with respect to some quasi-split torus, so $(\mathbf{T}, w_0F')$ is geometrically conjugate to $(\mathbf{T}_0, F)$ where $\mathbf{T}_0$ is a quasi-split torus. The $\mathbb{F}_q$-rank of $\mathbf{U}_n$ is thus the dimension of $(X(\mathbf{T}) \otimes \mathbb{R})^{w_0F'/q}$, *i.e.*, the number of eigenvalues q of w_0F', which is equal to $[n/2]$. Thus $\mathbf{U}_n$ is not split (and thus not isomorphic to $\mathbf{GL}_n$).

15.2 The symplectic group.

Let V be a $2n$-dimensional vector space over some algebraically closed field k. Let $(e_1,\ldots,e_n,e'_n,\ldots,e'_1)$ be a basis of V. The group $\mathbf{Sp}_{2n}$ is the group of automorphisms of $(V, \langle\, ,\, \rangle)$ where $\langle\, ,\, \rangle$ is the symplectic form given by $\langle e_i, e_j \rangle = \langle e'_i, e'_j \rangle = 0$ and $\langle e_i, e'_j \rangle = \delta_{i,j}$. It is clearly a closed subgroup of $\mathbf{GL}_{2n}$.

Maximal tori. A diagonal matrix is symplectic if and only if it is of the form

$$\begin{pmatrix} t_1 &&&&&&\\ & t_2 &&&&&\\ && \ddots &&&&\\ &&& t_n &&&\\ &&&& t_n^{-1} &&\\ &&&&& \ddots &\\ &&&&&& t_2^{-1} &\\ &&&&&&& t_1^{-1} \end{pmatrix}.$$

We will again let $\mathrm{diag}(t_1,\ldots,t_n)$ denote the above matrix. The set of such matrices is clearly a torus $\mathbf{T}$; this torus is maximal in $\mathbf{Sp}_{2n}$, as any maximal torus of $\mathbf{Sp}_{2n}$ is contained in some maximal torus of $\mathbf{GL}_{2n}$, and the only maximal torus of $\mathbf{GL}_{2n}$ containing $\mathbf{T}$ is the torus of all diagonal matrices.

Borel subgroups. Let $\mathbf{B}$ be the group of all symplectic upper triangular matrices. A matrix of the form

$$\begin{pmatrix} x & y \\ 0 & x' \end{pmatrix}$$

is symplectic if and only if ${}^t x' J y = {}^t y J x'$ and ${}^t x' J x = J$, where J is the matrix

$$\begin{pmatrix} & & 1 \\ & \cdot^{\cdot^{\cdot}} & \\ 1 & & \end{pmatrix}.$$

We get all the elements of $\mathbf{B}$ by taking an arbitrary triangular x and an arbitrary symmetric s, and putting $x' = J {}^t x^{-1} J$ and $y = J {}^t x'^{-1} s = xJs$. Thus $\mathbf{B}$ is isomorphic to the variety of $n \times n$ triangular matrices times the variety of $n \times n$ symmetric matrices and is thus connected. As it is the intersection of a Borel subgroup of $\mathbf{GL}_{2n}$ with $\mathbf{Sp}_{2n}$, it is solvable. Again we see that it is maximal since the only Borel subgroup of $\mathbf{GL}_{2n}$ which contains $\mathbf{B}$ is the Borel subgroup of all upper triangular matrices (as there is only one complete flag stabilized by all matrices in $\mathbf{B}$). So $\mathbf{B}$ is a Borel subgroup of $\mathbf{Sp}_{2n}$, which contains the maximal torus $\mathbf{T}$. Its unipotent radical $\mathbf{U}$ is the subgroup of unipotent matrices, *i.e.*, the group of all matrices

$$\begin{pmatrix} x & xJs \\ 0 & J {}^t x^{-1} J \end{pmatrix}$$

where s is symmetric and x upper triangular and unipotent. The group $\mathbf{B}$ is also the stabilizer of the complete (*i.e.*, maximal) flag of isotropic subspaces $<e_1> \subset <e_1, e_2> \subset \ldots \subset <e_1, \ldots, e_n>$, and by conjugating in $\mathbf{Sp}_{2n}$ we get other Borel subgroups as stabilizers of other complete isotropic flags.

As in $\mathbf{GL}_n$, we see that $\mathbf{Sp}_{2n}$ is reductive by observing that the intersection of the Borel subgroup of all symplectic lower triangular matrices with $\mathbf{B}$ is the maximal torus $\mathbf{T}$. The centre of $\mathbf{Sp}_{2n}$ is the set of scalar symplectic matrices which consists only of Id and $-$Id. Thus the radical is trivial and $\mathbf{Sp}_{2n}$ is semi-simple. Finally $\mathbf{Sp}_{2n}$ is connected: to see that it is enough to see that every element is in a Borel subgroup. Any $g \in \mathbf{Sp}_{2n}$ has at least one eigenvector $x \in V$; as g is symplectic, it induces a symplectic automorphism h of $V_1 = <x>^\perp/<x>$ (which is naturally endowed with a symplectic form induced by the initial form on V). By induction on $\dim V$, we may assume that $\mathbf{Sp}(V_1)$ is connected, and thus h is in some Borel subgroup of $\mathbf{Sp}(V_1)$, *i.e.*, stabilizes a complete isotropic flag of V_1. The inverse image in V of this flag, completed by $<x>$, is a complete isotropic flag of V stabilized by g, whence the result.

Roots. We denote by i, j elements in $1, \ldots, n$ and by i' and j' the corresponding elements in the second set of indices $1', \ldots, n'$. We denote by $\mathbf{U}_{i,j}$ (resp. $\mathbf{U}_{i,j'}$, resp. $\mathbf{U}_{i,i'}$) the group of unipotent matrices whose only non-zero off-diagonal entries are in positions (i, j) and (i', j') and have opposed values (resp. whose only non-zero off-diagonal entries are in positions (i, j') and (j, i') and have equal values, resp. whose only non-zero off-diagonal entry is in position (i, i')). These are all subgroups of $\mathbf{U}$ isomorphic to $\mathbb{G}_a$, normalized by $\mathbf{T}$; the element $\mathrm{diag}(t_1, \ldots, t_n) \in \mathbf{T}$ acts by multiplication by t_i/t_j (resp. $t_i t_j$, resp. t_i^2). We thus get n^2 distinct root subgroups. They represent all the positive roots since by the above description $\mathbf{U}$ is also of dimension n^2. The simple roots are the α_i (where $1 \leq i \leq n-1$) which map $\mathrm{diag}(t_1, \ldots, t_n)$ to t_i/t_{i+1}, and α_n, which maps $\mathrm{diag}(t_1, \ldots, t_n)$ to t_n^2. The root system thus obtained is of type C_n.

Parabolic subgroups. As the stabilizer $\mathbf{B}$ of any complete isotropic flag $V_1 \subset \ldots \subset V_n$ in $\mathbf{Sp}_{2n}$ is a Borel subgroup, the stabilizer of any subflag is a parabolic subgroup. We thus get 2^n distinct parabolic subgroups containing $\mathbf{B}$. Since there are also 2^n subsets of the set of simple roots, they are the only parabolic subgroups containing $\mathbf{B}$. As any isotropic flag may be completed to a complete one, we get the result that in general parabolic groups are the stabilizers of (complete or not) isotropic flags.

15.3 The split and non-split even-dimensional orthogonal groups. Orthogonal groups may be analyzed on the same lines as the symplectic groups, replacing the symplectic form by a quadratic. We may assume that the matrix of the quadratic form has the form $\begin{pmatrix} & & 1 \\ & \cdot^{\cdot^{\cdot}} & \\ 1 & & \end{pmatrix}$. In even-dimensional spaces, we will show that these groups have two possible rational structures, one of which is non-split.

Let V be a $2n$-dimensional vector space over some algebraically closed field k whose characteristic is different from 2. Let $(e_1, \ldots, e_n, e'_n, \ldots, e'_1)$ a basis of V; we endow V with the quadratic form defined by $\langle e_i, e_j \rangle = \langle e'_i, e'_j \rangle = 0$ and $\langle e_i, e'_j \rangle = \delta_{i,j}$; we will denote by Φ the matrix of this form. The group $\mathbf{O}_{2n}$ is the group of automorphisms of $(V, \langle\ ,\ \rangle)$. It is clearly a closed subgroup of $\mathbf{GL}_n(k) = \mathbf{GL}(V)$. The subgroup $\mathbf{SO}_{2n}$ of the elements of determinant 1 is a closed subgroup of index 2. Using the same arguments as for the symplectic group it may be shown that it is connected, and is thus the identity component of $\mathbf{O}_{2n}$.

Using the same arguments as for the symplectic groups, it may also be seen that the subgroup $\mathbf{T}$ of matrices of the form

$$\begin{pmatrix} t_1 &&&&&&& \\ & t_2 &&&&&& \\ && \ddots &&&&& \\ &&& t_n &&&& \\ &&&& t_n^{-1} &&& \\ &&&&& \ddots && \\ &&&&&& t_2^{-1} & \\ &&&&&&& t_1^{-1} \end{pmatrix}$$

is a maximal torus. We will denote the above element of $\mathbf{T}$ by $\operatorname{diag}(t_1, \ldots, t_n)$. Similarly we get a Borel subgroup $\mathbf{B}$ by taking the group of matrices of the form $\begin{pmatrix} x & xJa \\ 0 & J{}^t x^{-1} J \end{pmatrix}$ where x is an invertible upper triangular $n \times n$-matrix, where a is antisymmetric and where J is as before. The same argument as in the symplectic case shows that the Borel subgroups are the stabilizers of maximal isotropic flags and that $\mathbf{SO}_{2n}$ is reductive. The unipotent radical $\mathbf{U}$ of $\mathbf{B}$ consists of matrices $\begin{pmatrix} x & xJa \\ 0 & J{}^t x^{-1} J \end{pmatrix} \in \mathbf{B}$ where x is unipotent. Its dimension is thus $n(n-1)$. The root subgroups are the groups $\mathbf{U}_{i,j}$ (resp. $\mathbf{U}_{i,j'}$) of matrices whose only non-zero off-diagonal coefficients are in positions (i,j) and (j',i') with $i<j$ (resp. (i,j') and (j,i')) and have opposed values. The element $\operatorname{diag}(t_1, \ldots, t_n) \in \mathbf{T}$ acts on $\mathbf{U}_{i,j}$ (resp. $\mathbf{U}_{i,j'}$) through the root t_i/t_j (resp. $t_i t_j$). We thus get a set of n^2-n distinct positive roots of $\mathbf{SO}_{2n}$ with respect to $\mathbf{T}$; it is the whole set of positive roots since the dimension of $\mathbf{U}$ is n^2-n. The simple roots are $\alpha_i(\operatorname{diag}(t_1, \ldots, t_n)) = t_i/t_{i+1}$ for $1 \le i \le n-1$ and $\alpha_n(\operatorname{diag}(t_1, \ldots, t_n)) = t_{n-1}t_n$. If $n \ge 2$ the root system thus obtained is of type D_n; the centre of the group is the intersection of the kernels of the roots, which is ± 1; so $\mathbf{SO}_{2n}$ is semi-simple.

Frobenius endomorphisms. We assume now that $k = \overline{\mathbb{F}}_q$ (with $p \neq 2$). We will consider two Frobenius endomorphisms on $\mathbf{O}_{2n}$:

- The standard Frobenius endomorphism F which raises all coefficients of a matrix to the q-th power. The group $\mathbf{O}_{2n}^F$ we thus get is the group of the form $\langle\, ,\rangle$ over $\mathbb{F}_q$.
- The Frobenius endomorphism εF which is the composite of F with $\operatorname{ad} \varepsilon$ for some F-fixed element $\varepsilon \in \mathbf{O}_{2n} - \mathbf{SO}_{2n}$. As two such endomorphisms differ by an element of $\mathbf{SO}_{2n}$, the groups of rational points they define are isomorphic (they are conjugate by an inverse image under the Lang

map of the element by which the Frobenius endomorphisms differ). We will take for ε the element

$$\begin{pmatrix} 1 &&&&&&& \\ & \ddots &&&&&& \\ && 1 &&&&& \\ &&& 0 & 1 &&& \\ &&& 1 & 0 &&& \\ &&&&& 1 && \\ &&&&&& \ddots & \\ &&&&&&& 1 \end{pmatrix}.$$

This element normalizes both $\mathbf{T}$ and $\mathbf{B}$, and induces the non-trivial automorphism of the root system of type D_n.

We will show that $\mathbf{O}_{2n}^{\varepsilon F}$ is the group of a quadratic form which is not equivalent over $\mathbb{F}_q$ to $\langle\, , \rangle$. Let $h \in \mathbf{GL}_{2n}$ be such that $h^{-1}.{}^F h = \varepsilon$; then $\operatorname{ad} h$ maps the action of εF to that of F (in particular, it maps $\mathbf{O}_{2n}^{\varepsilon F}$ into $\mathbf{GL}_{2n}^F$). If g preserves the form Φ then ${}^h g$ preserves the form $\Phi_1 = {}^t h^{-1} \Phi h^{-1}$. As Φ is εF-stable, Φ_1 is F-stable. Thus $\operatorname{ad} h$ is an isomorphism from $\mathbf{O}_{2n}^{\varepsilon F}$ to the orthogonal group over $\mathbb{F}_q$ of Φ_1. Let us show that the quadratic forms Φ and Φ_1 are not equivalent under $\mathbf{GL}_{2n}(\mathbb{F}_q)$. If $a \in \mathbf{GL}_n(\mathbb{F}_q)$ maps Φ_1 to Φ, then ah is in the orthogonal group of Φ. But $(ah)^{-1}.{}^F(ah) = \varepsilon$, so this would imply $\det(\varepsilon) = \det(ah)^{q-1} = (\pm 1)^{q-1} = 1$, which is false, since $\det(\varepsilon) = -1$, whence the result. Looking at the action on $\mathbf{T}$, which is quasi-split since $\mathbf{B}$ is fixed by both F and εF, we notice that $\mathbf{O}_{2n}$ is split for F but not for εF.

15.4 The irreducible characters of $\mathbf{GL}_n(\mathbb{F}_q)$ and $\mathbf{U}_n(\mathbb{F}_q)$

We will express all irreducible characters of the linear and unitary groups as explicit combinations of Deligne-Lusztig characters, following the method of [G. LUSZTIG and B. SRINIVASAN The characters of the finite unitary groups, *Journal of algebra*, **49** (1977), 167–171]. Note that this will prove in particular that all class functions are uniform in these groups. We will denote by $\mathbf{G}$ the group $\mathbf{GL}_n$ and by F a Frobenius endomorphism which may be either the standard one or that of the unitary group (called F' in 15.1). We denote by $\mathbf{T}$ the torus of diagonal matrices. This torus is F-stable in both cases; it is split for linear groups, but is in no F-stable Borel subgroup in the unitary case. The Weyl group $W = N_{\mathbf{G}}(\mathbf{T})$ can be identified with the symmetric group, where $w \in W$ maps $\operatorname{diag}(t_1, t_2, \ldots, t_n) \in \mathbf{T}$ to $\operatorname{diag}(t_{w(1)}, \ldots, t_{w(n)})$ (see 15.1). In both cases the action of F on W is trivial (which is why we prefer to use $\mathbf{T}$ in the unitary case, rather than some quasi-split torus, as a reference to measure types of other tori). Thus classes of tori are parametrized by the conjugacy classes of W. For any $w \in W$ we put

$R_w = R^{\mathbf{G}}_{\mathbf{T}_w}(\mathrm{Id}_{\mathbf{T}_w})$, where $\mathbf{T}_w$ is a maximal torus of type w with respect to $\mathbf{T}$. The first step is to express all unipotent characters as linear combinations of the R_w. Let π_w be the normalized characteristic function of the class of $w \in W$ (see 12.19).

15.5 LEMMA. *The map $\pi_w \mapsto R_w$ induces an isometry from the space of class functions on W onto a subspace of the class functions on $\mathbf{G}^F$.*

PROOF: This follows immediately from 11.16. ■

We note that this lemma remains true in any reductive group, if we replace the classes of W in the above by the F-classes of W.

The image of $\chi \in \mathrm{Irr}(W)$ under the above isometry is the function

$$R_\chi = |W|^{-1} \sum_{w \in W} \chi(w) R_w.$$

The set of functions $\{ R_\chi \mid \chi \in \mathrm{Irr}(W) \}$ is thus orthonormal.

15.6 REMARK. By 12.13, we have $R_{\mathrm{Id}_W} = \mathrm{Id}_{\mathbf{G}^F}$, and as observed in the proof of 12.14 we have $R_{\mathrm{sgn}} = \mathrm{St}_{\mathbf{G}}$, where sgn is the "sign" character $(-1)^{l(w)}$ of W.

We will show that in the case of linear and unitary groups the set $\{ R_\chi \mid \chi \in \mathrm{Irr}(W) \}$ is actually equal (up to sign in the latter case) to the set of unipotent characters $\mathcal{E}(\mathbf{G}^F, (1))$; in other groups some of the R_χ are only combinations of irreducible characters with coefficients in $\mathbb{Q}$. We will let $R^{\mathbf{G}}_w$, $R^{\mathbf{G}}_\chi$, $\pi^{\mathbf{G}}_w$, $W_{\mathbf{G}}$ *etc.* denote the above objects whenever there may be any ambiguity on the group considered. We note that these objects are well-defined in any reductive group once a reference torus $\mathbf{T}$ has been chosen. We shall extend by linearity the notation R_χ to any class function.

Let $\mathbf{L}$ be a "diagonal" Levi subgroup of $\mathbf{G}$, *i.e.*, a Levi subgroup which consists of block-diagonal matrices. Such a Levi subgroup is F-stable and contains $\mathbf{T}$. It is isomorphic in the linear case to a product of linear groups and in the unitary case to a product of unitary groups. The next proposition shows that $R^{\mathbf{G}}_{\mathbf{L}}$ is "transported" from ordinary induction in the Weyl group.

15.7 PROPOSITION. *For any class function f on $W_{\mathbf{L}}$ we have $R^{\mathbf{G}}_{\mathbf{L}}(R_f) = R_{\mathrm{Ind}^{W_{\mathbf{G}}}_{W_{\mathbf{L}}}(f)}$.*

PROOF: It is enough to prove the proposition for the functions $\pi^{\mathbf{L}}_w$. But $\mathrm{Ind}^{W_{\mathbf{G}}}_{W_{\mathbf{L}}}(\pi^{\mathbf{L}}_w) = \pi^{\mathbf{G}}_w$, and by the isometry 15.5 the image of $\pi^{\mathbf{L}}_w$ (resp. $\pi^{\mathbf{G}}_w$) is $R^{\mathbf{L}}_w$ (resp. $R^{\mathbf{G}}_w$), whence the result since $R^{\mathbf{G}}_{\mathbf{L}}(R^{\mathbf{L}}_w) = R^{\mathbf{G}}_w$ (see 11.5). ■

We may now prove

15.8 THEOREM. *In the linear and unitary groups, the characters $\{R_\chi\}_{\chi\in\mathrm{Irr}(W)}$ are (up to sign in the unitary case) the unipotent characters.*

PROOF: If we prove that the R_χ are virtual characters, they will be up to sign distinct irreducible characters (since they form an orthonormal set) and, as $R_w = \sum_\chi \chi(w)R_\chi$, they will span the unipotent characters.

We know that R_{Id} is an irreducible character (the character $\mathrm{Id}_{\mathbf{G}}$). It is also well-known that in the symmetric group any irreducible character is a linear combination with integral coefficients of the identity induced from various parabolic subgroups *i.e.*, any $\chi \in \mathrm{Irr}(W)$ is a linear combination with integral coefficients of $\mathrm{Ind}_{W_{\mathbf{L}}}^{W}(\mathrm{Id})$ where $\mathbf{L}$ runs over diagonal Levi subgroups of $\mathbf{G}$. Applying the isometry 15.5 and proposition 15.7, we get an expression of R_χ as an integral linear combination of $R_{\mathbf{L}}^{\mathbf{G}}(\mathrm{Id})$, which proves indeed that it is a virtual character.

To prove that the R_χ are actual characters for linear groups, we note that, in that case, as the torus $\mathbf{T}$ is split it is contained in a rational Borel subgroup $\mathbf{B}$ (*e.g.*, the Borel subgroup of all upper triangular matrices), thus $R_1 = R_{\mathbf{T}}^{\mathbf{G}}(\mathrm{Id}_{\mathbf{T}}) = \mathrm{Ind}_{\mathbf{B}^F}^{\mathbf{G}^F}(\mathrm{Id}_{\mathbf{B}})$ is an actual character; since $R_1 = \sum_\chi \chi(1)R_\chi$, and $\chi(1) > 0$, this proves the result. ■

In the unitary case, it is not difficult to obtain a formula for the dimension of R_χ and deduce from it, for each χ, which of R_χ or $-R_\chi$ is an irreducible character. Note that, as $\mathbf{T}$ is of type w_0 with respect to some quasi-split torus (see description of the unitary groups at the end of 15.1), the same argument as in the linear case shows that R_{w_0} is an actual character, so $\chi(w_0)R_\chi$ also. So, when $\chi(w_0) \neq 0$, the sign we have to give to R_χ to make it an actual character is that of $\chi(w_0)$.

15.9 The character tables of $\mathbf{GL}_2(\mathbb{F}_q)$ and $\mathbf{SL}_2(\mathbb{F}_q)$
To determine the character tables of these groups, we first list their irreducible characters by applying the above results.

We will denote the two elements of the Weyl group $W(\mathbf{T})$ as $\{1, s\}$, where s is the transposition $(1,2)$. The group of rational points $\mathbf{T}^F$ of the diagonal torus is isomorphic to $\mathbb{F}_q^\times \times \mathbb{F}_q^\times$ in $\mathbf{GL}_2$ (resp. to the group μ_{q-1} of $(q-1)$-th roots of 1 in the case of $\mathbf{SL}_2$). Thus an element of $\mathrm{Irr}(\mathbf{T}^F)$ is given by a pair (α, β) of characters of $\mathbb{F}_q^\times$ (resp. a character α of μ_{q-1}), and s acts by exchanging α and β (resp. sending α to its inverse). By 11.15, the Deligne-Lusztig characters $R^{\mathbf{G}}_{\mathbf{T}}(\alpha,\beta)$ (resp. $R^{\mathbf{G}}_{\mathbf{T}_s}(\alpha)$) are all distinct when $\{\alpha,\beta\}$ runs over (non-ordered) pairs of characters of $\mathbb{F}_q^\times$ (resp. representatives of $\mathrm{Irr}(\mu_{q-1})$ mod $\alpha\equiv\alpha^{-1}$), and are irreducible (see remark after 13.1) when $\alpha\neq\beta$ (resp. $\alpha^2\neq 1$). When they are not irreducible, since by 11.15 their norm is 2, they have two distinct irreducible components.

We choose a torus $\mathbf{T}_s$ of type s with respect to the diagonal torus $\mathbf{T}$. We have

$$\mathbf{T}_s^F\simeq\mathbf{T}^{sF}=\{\,\mathrm{diag}(t_1,t_1^q)\mid t_1\in\mathbb{F}_{q^2}^\times\ (\text{resp. } t_1^{q+1}=1)\,\}.$$

Thus $\mathbf{T}_s^F\simeq\mathbb{F}_{q^2}^\times$ (resp. μ_{q+1}). A character of $\mathbf{T}_s^F$ is given by some character ω of $\mathbb{F}_{q^2}^\times$ (resp. of μ_{q+1}), and s acts by $\omega\mapsto\omega^q$ (and $\omega^q=\omega^{-1}$ in the case of $\mathbf{SL}_2$). Thus the Deligne-Lusztig characters $R^{\mathbf{G}}_{\mathbf{T}_s}(\omega)$ are all distinct when ω runs over representatives of $\mathrm{Irr}(\mathbb{F}_{q^2}^\times)$ (resp. of $\mathrm{Irr}(\mu_{q+1})$) mod $\omega\equiv\omega^q$; these characters are (the opposite by 12.10 of) irreducible characters unless $\omega=\omega^q$ (which in the case of $\mathbf{SL}_2$ implies that q is odd and $\omega^2=1$); and when they are not irreducible they have two irreducible constituents. When they are irreducible the characters $R^{\mathbf{G}}_{\mathbf{T}}(\alpha,\beta)$ and $-R^{\mathbf{G}}_{\mathbf{T}_s}(\omega)$ are all distinct since they are in different Lusztig series.

We now decompose the non-irreducible $R^{\mathbf{G}}_{\mathbf{T}}(\theta)$. In the case of $\mathbf{GL}_2$, using the results of 15.4, we have $R^{\mathbf{G}}_{\mathbf{T}}(\mathrm{Id}_{\mathbf{T}})=\mathrm{Id}_{\mathbf{G}}+\mathrm{St}_{\mathbf{G}}$ and $R^{\mathbf{G}}_{\mathbf{T}_s}(\mathrm{Id}_{\mathbf{T}_s})=\mathrm{Id}_{\mathbf{G}}-\mathrm{St}_{\mathbf{G}}$. To decompose the other $R^{\mathbf{G}}_{\mathbf{T}}(\alpha,\alpha)$ we may use the bijection of 13.30 (ii). In the case of $\mathbf{GL}_n$ the linear characters are of the form $\alpha\circ\det$. We have

$$R^{\mathbf{G}}_{\mathbf{T}}(\alpha,\alpha)=R^{\mathbf{G}}_{\mathbf{T}}(\mathrm{Id}_{\mathbf{T}}).(\alpha\circ\det)=\mathrm{Id}_{\mathbf{G}}.(\alpha\circ\det)+\mathrm{St}_{\mathbf{G}}.(\alpha\circ\det).$$

Similarly, when $\omega\in\mathrm{Irr}(\mathbb{F}_{q^2}^\times)$ is of order $q-1$, there exists $\alpha\in\mathrm{Irr}(\mathbb{F}_q{}^\times)$ such that $\omega=\alpha\circ N_{\mathbb{F}_{q^2}/\mathbb{F}_q}$ (and ω is the restriction to $\mathbf{T}_s^F$ of the linear character $\alpha\circ\det$ of $\mathbf{G}^F$); we have

$$R^{\mathbf{G}}_{\mathbf{T}_s}(\omega)=\mathrm{Id}_{\mathbf{G}}.(\alpha\circ\det)-\mathrm{St}_{\mathbf{G}}.(\alpha\circ\det).$$

We consider now $\mathbf{SL}_2$. As observed above, if the characteristic is 2 all Deligne-Lusztig characters except $R^{\mathbf{G}}_{\mathbf{T}}(\mathrm{Id}_{\mathbf{T}})$ and $R^{\mathbf{G}}_{\mathbf{T}_s}(\mathrm{Id}_{\mathbf{T}_s})$ are irreducible. If q is odd, we must decompose $R^{\mathbf{G}}_{\mathbf{T}}(\alpha_0)$, where α_0 is the character of order 2 of μ_{q-1} and $R^{\mathbf{G}}_{\mathbf{T}_s}(\omega_0)$, where ω_0 is the character of order 2 of μ_{q+1}. By 13.22, we have

$$R^{\mathbf{SL}_2}_{\mathbf{T}\cap\mathbf{SL}_2}(\mathrm{Res}^{\mathbf{T}^F}_{\mathbf{T}^F\cap\mathbf{SL}_2}\theta)=\mathrm{Res}^{\mathbf{GL}_2^F}_{\mathbf{SL}_2^F}(R^{\mathbf{GL}_2}_{\mathbf{T}}(\theta)),$$

so, being the restriction of an (actual) irreducible character, $R^{\mathbf{G}}_{\mathbf{T}}(\alpha_0)$ is the sum of two distinct irreducible characters. We put $R^{\mathbf{G}}_{\mathbf{T}}(\alpha_0)=\chi^+_{\alpha_0}+\chi^-_{\alpha_0}$. Similarly we may put $-R^{\mathbf{G}}_{\mathbf{T}_s}(\omega_0)=\chi^+_{\omega_0}+\chi^-_{\omega_0}$; the characters $\chi^\pm_{\alpha_0}$ must be distinct from $\chi^{\pm}_{\omega_0}$, since $R^{\mathbf{G}}_{\mathbf{T}_s}(\omega_0)$ and $R^{\mathbf{G}}_{\mathbf{T}}(\alpha_0)$ are orthogonal. Actually $R^{\mathbf{G}}_{\mathbf{T}}(\alpha_0)=\chi_{(s_1)}$ and $R^{\mathbf{G}}_{\mathbf{T}_s}(\omega_0)=\chi_{(s_2)}$ where s_1 is the element of order 2 of the split torus of $\mathbf{PGL}_2$ and s_2 is the element of order 2 of the non-split torus of $\mathbf{PGL}_2$, and, since s_1 and s_2 are not rationally conjugate, they are in distinct rational series (see 14.40 and below).

The list of irreducible characters of $\mathbf{GL}_2(\mathbb{F}_q)$ is thus

- The $R^{\mathbf{G}}_{\mathbf{T}}(\alpha,\beta)$ where (α,β) is an (unordered) pair of distinct characters of $\mathbb{F}_q^\times$.
- The characters $\mathrm{Id}_{\mathbf{G}}.(\alpha\circ\det)$ and $\mathrm{St}_{\mathbf{G}}.(\alpha\circ\det)$ where $\alpha\in\mathrm{Irr}(\mathbb{F}_q^\times)$.
- The $-R^{\mathbf{G}}_{\mathbf{T}_s}(\omega)$ where $\omega\in\mathrm{Irr}(\mathbb{F}_{q^2}^\times)$ is such that $\omega^q\neq\omega$ (where $-R^{\mathbf{G}}_{\mathbf{T}_s}(\omega)=R^{\mathbf{G}}_{\mathbf{T}_s}(\omega^q)$).

And the list of irreducible characters of $\mathbf{SL}_2(\mathbb{F}_q)$ is

- The $R^{\mathbf{G}}_{\mathbf{T}}(\alpha)$ where $\alpha\in\mathrm{Irr}(\mathbb{F}_q^\times)$ is such that $\alpha^2\neq 1$ (and where $R^{\mathbf{G}}_{\mathbf{T}}(\alpha)=R^{\mathbf{G}}_{\mathbf{T}}(\alpha^{-1})$).
- The characters $\mathrm{Id}_{\mathbf{G}}$ and $\mathrm{St}_{\mathbf{G}}$.
- If q is odd, the two irreducible constituents $\chi^+_{\alpha_0}$ and $\chi^-_{\alpha_0}$ of $R^{\mathbf{G}}_{\mathbf{T}}(\alpha_0)$.
- The $-R^{\mathbf{G}}_{\mathbf{T}_s}(\omega)$ where $\omega\in\mu_{q+1}$ is such that $\omega^2\neq 1$ (and where $R^{\mathbf{G}}_{\mathbf{T}_s}(\omega)=R^{\mathbf{G}}_{\mathbf{T}_s}(\omega^{-1})$).
- If q is odd, the two irreducible constituents $\chi^+_{\omega_0}$ and $\chi^-_{\omega_0}$ of $-R^{\mathbf{G}}_{\mathbf{T}_s}(\omega_0)$.

We now give the list of conjugacy classes of $\mathbf{GL}_2(\mathbb{F}_q)$ and $\mathbf{SL}_2(\mathbb{F}_q)$. In general, to list the conjugacy classes of a connected reductive group, we may first describe the semi-simple classes and their centralizers, and then for each such centralizer $C_{\mathbf{G}}(s)$ describe the unipotent classes of $C_{\mathbf{G}}(s)$ (which are actually in $C^\circ_{\mathbf{G}}(s)$ by 2.5) and their centralizers. We then get the classes of $\mathbf{G}^F$ by 3.15 (iv) and 3.25. By 2.6 (i), the semi-simple classes of $\mathbf{GL}_n^F$ (resp. $\mathbf{SL}_n^F$) are in one-to-one correspondence with the rational semi-simple classes of $\mathbf{GL}_n$ (resp. $\mathbf{SL}_n$); in $\mathbf{GL}_2$, the only non-semi-simple classes are of the form zu with $z\in Z(\mathbf{GL}_2)$ and u a non-trivial unipotent element: they have as representative a matrix of the form $\begin{pmatrix} a & b\\ 0 & a\end{pmatrix}$ (when $a=1$ they have been analysed in 3.26 (ii)).

We list first the semi-simple classes; by 3.26 (ii) and 2.6 (i), the semi-simple classes of $\mathbf{SL}_n^F$ are the semi-simple classes of $\mathbf{GL}_n^F$ which lie in $\mathbf{SL}_n$. In $\mathbf{GL}_n^F$ the semi-simple elements of $\mathbf{T}_w^F\simeq\mathbf{T}^{wF}$ may be identified with matrices $\mathrm{diag}(t_1,\ldots,t_n)$ such that $t_{w(i)}=t_i^q$. Two such elements are conjugate in $\mathbf{GL}_n$ if and only if they are conjugate under W (see 0.9). The list we get in $\mathbf{GL}_2$ is

- From $\mathbf{T}^F$, we get $Z(\mathbf{G})^F$, whose elements $\begin{pmatrix} a & 0 \\ 0 & a \end{pmatrix}$ with $a \in \mathbb{F}_q^\times$ are all in distinct classes (these classes are in $\mathbf{SL}_2$ if $a \in \{-1, 1\}$); and the classes of $\begin{pmatrix} a & 0 \\ 0 & b \end{pmatrix}$ with $a, b \in \mathbb{F}_q^\times$ and $a \neq b$ (which are in $\mathbf{SL}_2$ if $a = b^{-1}$).
- From $\mathbf{T}_s^F$, we get the classes of $\begin{pmatrix} x & 0 \\ 0 & {}^F x \end{pmatrix}$ with $x \in \mathbb{F}_{q^2}^\times$; these classes do not meet $\mathbf{T}^F$ if $x \notin \mathbb{F}_q$, *i.e.*, $x \neq {}^F x$, which is also equivalent to $\begin{pmatrix} x & 0 \\ 0 & {}^F x \end{pmatrix} \notin Z(\mathbf{G})$. A form with rational coefficients of these elements is $\begin{pmatrix} 0 & 1 \\ -x.{}^F x & x + {}^F x \end{pmatrix}$; these classes are in $\mathbf{SL}_2$ if $x.{}^F x = 1$.

We now list those classes whose centralizer is not a torus. In $\mathbf{GL}_2$, as remarked above, these are the classes zu where z is central and u unipotent. As analysed in 3.26 (ii) there is in $\mathbf{GL}_2(\mathbb{F}_q)$ only one class of non-trivial unipotent elements, and in $\mathbf{SL}_2(\mathbb{F}_q)$ two classes if the characteristic is different from 2 (these unipotent elements are regular so we could also take the analysis from 14.24). We get in $\mathbf{GL}_2(\mathbb{F}_q)$ the classes of $\begin{pmatrix} a & 1 \\ 0 & a \end{pmatrix}$ with $a \in \mathbb{F}_q^\times$, and in $\mathbf{SL}_2(\mathbb{F}_q)$, when the characteristic is different from 2, the four classes $\begin{pmatrix} a & b \\ 0 & a \end{pmatrix}$ with $a \in \{-1, 1\}$ and $b \in (\mathbb{F}_q^\times)^2$ or $b \in \mathbb{F}_q^\times - (\mathbb{F}_q^\times)^2$ (these are only one class in characteristic 2).

We now describe how to obtain the character tables of $\mathbf{GL}_2(\mathbb{F}_q)$ and of $\mathbf{SL}_2(\mathbb{F}_q)$ given in table 1 and table 2. All the entries in the character table of $\mathbf{GL}_2(\mathbb{F}_q)$ may be obtained by applying the character formula 12.2, once we know the values of the Green functions $R_{\mathbf{T}}^{\mathbf{G}}(\mathrm{Id}_{\mathbf{T}})$ and $R_{\mathbf{T}_s}^{\mathbf{G}}(\mathrm{Id}_{\mathbf{T}_s})$. These last values are known from $R_{\mathbf{T}}^{\mathbf{G}}(\mathrm{Id}_{\mathbf{T}}) = \mathrm{Id}_{\mathbf{G}} + \mathrm{St}_{\mathbf{G}}$ and $R_{\mathbf{T}_s}^{\mathbf{G}}(\mathrm{Id}_{\mathbf{T}_s}) = \mathrm{Id}_{\mathbf{G}} - \mathrm{St}_{\mathbf{G}}$, where the values of $\mathrm{St}_{\mathbf{G}}$ are given by 9.3. This also gives by restriction the complete table of $\mathbf{SL}_2(\mathbb{F}_q)$ in characteristic 2; in table 2 we assume that q is odd.

Most of the table of $\mathbf{SL}_2$ (q odd) is obtained by restriction from $\mathbf{GL}_2$ (using 13.22); the only values which are not given by restriction are those of $\chi_{\alpha_0}^\pm$ and $\chi_{\omega_0}^\pm$. Since $\chi_{\alpha_0}^+ + \chi_{\alpha_0}^-$ is the restriction of an irreducible character of $\mathbf{GL}_2(\mathbb{F}_q)$, the characters $\chi_{\alpha_0}^+$ and $\chi_{\alpha_0}^-$ are conjugate under $\mathbf{GL}_2(\mathbb{F}_q)$ (because $\mathbf{SL}_2(\mathbb{F}_q)$ is a normal subgroup of $\mathbf{GL}_2(\mathbb{F}_q)$). They have thus the same dimension, and more generally the same value on all classes of $\mathbf{SL}_2(\mathbb{F}_q)$ that are invariant under the action of $\mathbf{GL}_2(\mathbb{F}_q)$; in particular they also have the same central character α_0. These remarks show that to get all their values it is enough

Table 1: characters of $\mathbf{GL}_2(\mathbb{F}_q)$
(note that $|\mathbf{GL}_2(\mathbb{F}_q)| = q(q-1)^2(q+1)$)

Classes	$\begin{pmatrix} a & 0 \\ 0 & a \end{pmatrix}$ $a \in \mathbb{F}_q^\times$	$\begin{pmatrix} a & 0 \\ 0 & b \end{pmatrix}$ $a, b \in \mathbb{F}_q^\times$ $a \neq b$	$\begin{pmatrix} x & 0 \\ 0 & {}^F x \end{pmatrix}$ $x \in \mathbb{F}_{q^2}^\times$ $x \neq {}^F x$	$\begin{pmatrix} a & 1 \\ 0 & a \end{pmatrix}$ $a \in \mathbb{F}_q^\times$
Number of classes of this type	$q-1$	$\frac{(q-1)(q-2)}{2}$	$\frac{q(q-1)}{2}$	$q-1$
Cardinal of the class	1	$q(q+1)$	$q(q-1)$	q^2-1
$R_\mathbf{T}^\mathbf{G}(\alpha,\beta)$ $\alpha,\beta \in \mathrm{Irr}(\mathbb{F}_q^\times)$ $\alpha \neq \beta$	$(q+1)\alpha(a)\beta(a)$	$\alpha(a)\beta(b)+$ $\alpha(b)\beta(a)$	0	$\alpha(a)\beta(a)$
$-R_{\mathbf{T}_s}^\mathbf{G}(\omega)$ $\omega \in \mathrm{Irr}(\mathbb{F}_{q^2}^\times)$ $\omega \neq \omega^q$	$(q-1)\omega(a)$	0	$-\omega(x)-\omega({}^F x)$	$-\omega(a)$
$\mathrm{Id}_\mathbf{G}.(\alpha \circ \det)$ $\alpha \in \mathrm{Irr}(\mathbb{F}_q^\times)$	$\alpha(a^2)$	$\alpha(ab)$	$\alpha(x.{}^F x)$	$\alpha(a^2)$
$\mathrm{St}_\mathbf{G}.(\alpha \circ \det)$ $\alpha \in \mathrm{Irr}(\mathbb{F}_q^\times)$	$q\alpha(a^2)$	$\alpha(ab)$	$-\alpha(x.{}^F x)$	0

to compute their values on the classes $\begin{pmatrix} 1 & b \\ 0 & 1 \end{pmatrix}$. The same analysis holds for $\chi_{\omega_0}^+$ and $\chi_{\omega_0}^-$. These classes are regular unipotent; if 1 and z are the two elements of $H^1(F, Z(\mathbf{SL}_2))$ and we choose for u_1 (see definition after 14.25) the class of $\begin{pmatrix} 1 & 1 \\ 0 & 1 \end{pmatrix}$, then u_z is the class of $\begin{pmatrix} 1 & x \\ 0 & 1 \end{pmatrix}$ with $x \in \mathbb{F}_q^\times - (\mathbb{F}_q^\times)^2$. For any $\chi \in \mathrm{Irr}(\mathbf{SL}_2(\mathbb{F}_q))$, by 14.35 (i), we have

$$\left.\begin{aligned} -\chi(u_1) &= \sigma_1\langle D_\mathbf{G}(\chi), \Gamma_1\rangle + \sigma_z\langle D_\mathbf{G}(\chi), \Gamma_z\rangle \\ \text{and } -\chi(u_z) &= \sigma_1\langle D_\mathbf{G}(\chi), \Gamma_z\rangle + \sigma_z\langle D_\mathbf{G}(\chi), \Gamma_1\rangle. \end{aligned}\right\} \quad (1)$$

As we have seen $\chi_{\alpha_0}^+$ and $\chi_{\alpha_0}^-$ are the two components of $\chi_{(s_1)}$ and $\chi_{\omega_0}^+$ and $\chi_{\omega_0}^-$ are the two components of $\chi_{(s_2)}$. Let us choose for $\chi_{\alpha_0}^+$ (resp. $\chi_{\omega_0}^+$) the common component of $\chi_{(s_1)}$ and Γ_1 (resp. of $\chi_{(s_2)}$ and Γ_1) (see 14.44). Since $D_\mathbf{G}(R_\mathbf{T}^\mathbf{G}(\alpha_0)) = R_\mathbf{T}^\mathbf{G}(\alpha_0)$ and $D_\mathbf{G}(R_{\mathbf{T}_s}^\mathbf{G}(\omega_0)) = -R_{\mathbf{T}_s}^\mathbf{G}(\omega_0)$, and since $\langle \Gamma_z, D_\mathbf{G}(\Gamma_1)\rangle = 0$ (see 14.33 (ii)) we must have $D_\mathbf{G}(\chi_{\alpha_0}^+) = \chi_{\alpha_0}^+$, $D_\mathbf{G}(\chi_{\alpha_0}^-) = \chi_{\alpha_0}^-$, $D_\mathbf{G}(\chi_{\omega_0}^+) = -\chi_{\omega_0}^+$ and $D_\mathbf{G}(\chi_{\omega_0}^-) = -\chi_{\omega_0}^-$, so (1) gives

$$\chi_{\alpha_0}^+(u_1) = \chi_{\alpha_0}^-(u_z) = -\chi_{\omega_0}^+(u_1) = -\chi_{\omega_0}^-(u_z) = -\sigma_1$$

and

$$\chi_{\alpha_0}^+(u_z) = \chi_{\alpha_0}^-(u_1) = -\chi_{\omega_0}^+(u_z) = -\chi_{\omega_0}^-(u_1) = -\sigma_z.$$

Table 2: characters of $\mathbf{SL}_2(\mathbb{F}_q)$ for q odd
(note that $|\mathbf{SL}_2(\mathbb{F}_q)| = q(q-1)(q+1)$)

Classes	$\begin{pmatrix} a & 0 \\ 0 & a \end{pmatrix}$ $a \in \{1,-1\}$	$\begin{pmatrix} a & 0 \\ 0 & a^{-1} \end{pmatrix}$ $a \in \mathbb{F}_q^\times$ $a \neq \{1,-1\}$	$\begin{pmatrix} x & 0 \\ 0 & {}^F x \end{pmatrix}$ $x.{}^F x = 1$ $x \neq {}^F x$	$\begin{pmatrix} a & b \\ 0 & a \end{pmatrix}$ $a \in \{1,-1\}$, $b \in \{1, x\}$ with $x \in \mathbb{F}_q^\times - (\mathbb{F}_q^\times)^2$
Number of classes of this type	2	$(q-3)/2$	$(q-1)/2$	4
Cardinal of the class	1	$q(q+1)$	$q(q-1)$	$(q^2-1)/2$
$R_{\mathbf{T}}^{\mathbf{G}}(\alpha)$ $\alpha \in \mathrm{Irr}(\mathbb{F}_q^\times)$ $\alpha^2 \neq \mathrm{Id}$	$(q+1)\alpha(a)$	$\alpha(a) + \alpha(\frac{1}{a})$	0	$\alpha(a)$
$\chi_{\alpha_0}^\varepsilon$ $\varepsilon \in \{1,-1\}$	$\frac{q+1}{2}\alpha_0(a)$	$\alpha_0(a)$	0	$\frac{\alpha_0(a)}{2}(1+$ $\varepsilon\alpha_0(ab)\sqrt{\alpha_0(-1)q})$
$-R_{\mathbf{T}_s}^{\mathbf{G}}(\omega)$ $\omega \in \mathrm{Irr}(\mu_{q+1})$ $\omega^2 \neq \mathrm{Id}$	$(q-1)\omega(a)$	0	$-\omega(x) - \omega({}^F x)$	$-\omega(a)$
$\chi_{\omega_0}^\varepsilon$ $\varepsilon \in \{1,-1\}$	$\frac{q-1}{2}\omega_0(a)$	0	$-\omega_0(x)$	$\frac{\omega_0(a)}{2}(-1+$ $\varepsilon\alpha_0(ab)\sqrt{\alpha_0(-1)q})$
$\mathrm{Id}_{\mathbf{G}}$	1	1	1	1
$\mathrm{St}_{\mathbf{G}}$	q	1	-1	0

To compute σ_1 and σ_z, we notice that if $\chi \in \mathrm{Irr}(\mathbb{F}_q{}^+)$ is a character as defined above 14.29, then by 14.34 and 14.36 we have

$$\sigma_z = |Z(\mathbf{G}^F)|^{-1}\Gamma_z(u_1) = |Z(\mathbf{G}^F)|^{-1}\Gamma_1(u_z) = \sum_{x \in \mathbb{F}_q - (\mathbb{F}_q)^2} \chi(x)$$

and

$$\sigma_1 = |Z(\mathbf{G}^F)|^{-1}\Gamma_1(u_1) = \sum_{x \in (\mathbb{F}_q{}^\times)^2} \chi(x).$$

To compute them notice that $\sigma_1 + \sigma_z = -1$ and that $\sigma_1 - \sigma_z$ is the Gauss sum

$$\sum_{x \in \mathbb{F}_q} \alpha_0(x)\chi(x),$$

so $(\sigma_1 - \sigma_z)^2 = \alpha_0(-1)q$.

BIBLIOGRAPHY

This bibliography includes only the most often needed references; others appear on-line in the text. A quasi-exhaustive bibliography up to 1986 can be found in [Ca].

[B1] A. BOREL *Linear algebraic groups*, Benjamin (1969).

[B2] A. BOREL *et al.* Seminar on algebraic groups and related topics, *Lecture Notes in Mathematics,* **131** (1970), Springer.

[Bbk] N. BOURBAKI *Groupes et algèbres de Lie*, IV, V, VI, Masson (1981).

[BT] A. BOREL & J. TITS Groupes réductifs, *Publications Mathématiques de l'IHES,* **27** (1965), 55–160.

[Ca] R. W. CARTER *Finite groups of Lie type*, Wiley-Interscience (1985).

[Cu] C. W. CURTIS Truncation and duality in the character ring of a finite group of Lie type, *Journal of Algebra,* **62** (1980), 320–332.

[CuR] C. W. CURTIS & I. REINER *Methods of representation theory I*, Wiley-Interscience (1981).

[DL1] P. DELIGNE & G. LUSZTIG Representations of reductive groups over finite fields, *Annals of Math.,* **103** (1976), 103–161.

[DL2] P. DELIGNE & G. LUSZTIG Duality for representations of a reductive group over a finite field, *J. of Algebra,* **74** (1982), 284–291; and Duality for representations of a reductive group over a finite field II, *J. of Algebra,* **81** (1983), 540–545.

[DM1] F. DIGNE & J. MICHEL Remarques sur la dualité de Curtis, *Journal of Algebra,* **79, 1** (1982), 151–160.

[DM2] F. DIGNE & J. MICHEL Foncteurs de Lusztig et caractères des groupes linéaires et unitaires sur un corps fini, *Journal of Algebra,* **107, 1** (1987), 217–255.

[Go] D. GORENSTEIN Finite groups, Chelsea Pub. (1980).

[Ha] R. HARTSHORNE Algebraic Geometry, (*Graduate Texts in Mathematics,* **52**), Springer (1977).

[HC] HARISH-CHANDRA Eisenstein series over finite fields, *Functional Analysis and Related Fields,* Springer (1970), 76–88.

[Hu] J. E. HUMPHREYS Linear algebraic groups (*Graduate texts in mathematics,* **21**), Springer (1975).

[L1] G. LUSZTIG On the finiteness of the number of unipotent classes, *Inventiones Math.,* **34** (1976), 201–213.

[L2] G. LUSZTIG Representations of finite Chevalley groups, *CBMS Regional Conference Series in Mathematics (AMS),* **39** (1977).

[L3] G. LUSZTIG Representations of finite classical groups, *Inventiones Math.,* **43** (1977), 125–175.

[L4] G. LUSZTIG Characters of reductive groups over a finite field, (*Annals of math. studies,* **107**), Princeton University Press (1984).

[Se] J.-P. SERRE *Représentations linéaires des groupes finis,* Hermann (1971).

[SGA4] M. ARTIN *et al.* SGA 4. Théorie des topos et cohomologie étale des schéma, *Lecture Notes in Mathematics,* **269, 270, 305** (1972–1973), Springer.

[SGA4$\frac{1}{2}$] P. DELIGNE SGA 4$\frac{1}{2}$. Cohomologie étale, *Lecture Notes in Mathematics,* **569** (1977), Springer.

[SGA5] A. GROTHENDIECK *et al.* SGA 5. Cohomologie l-adique et fonctions L (Ed. L. Illusie), *Lecture Notes in Mathematics,* **589** (1977), Springer.

[Sp] T. A. SPRINGER *Linear algebraic groups,* Birkhäuser (1981).

[Sr] B. SRINIVASAN Representations of finite Chevalley groups, *Lecture notes in mathematics,* **764** (1979), Springer.

[St1] R. STEINBERG Endomorphisms of linear algebraic groups, *Memoirs of the American Mathematical Society,* **80** (1968).

[St2] R. STEINBERG Conjugacy classes in algebraic groups, *Lecture notes in mathematics,* **366** (1974), Springer

[Z] A. ZHELEVINSKI Representations of finite classical groups, *Lecture notes in mathematics,* **869** (1981), Springer.

INDEX